LOBSTERS

BIOLOGY, BEHAVIOR AND MANAGEMENT

MARINE AND FRESHWATER BIOLOGY

MARINE AND FRESHWATER BIOLOGY

LOBSTERS

BIOLOGY, BEHAVIOR AND MANAGEMENT

BRADY K. QUINN
EDITOR

nova
science publishers
New York

NOTICE TO THE READER

Library of Congress Cataloging-in-Publication Data

ISBN: 978-1-53615-711-6

Published by Nova Science Publishers, Inc. † New York

CONTENTS

Preface   vii

Chapter 1   Estimating Intrinsic Optimum Temperatures
             and Lower and Upper Thermal Thresholds for the
             Development of American Lobster Larvae
             Using a Thermodynamic Model   1
             Brady K. Quinn

Chapter 2   The Cryptic *Homarus gammarus* (L., 1758)
             Juveniles: A Comparative Approach to the
             Mystery of Their Whereabouts   61
             Gro I. van der Meeren and Astrid K. Woll

Chapter 3   Biologic and Socioeconomic Harvesting Strategies
             for the Caribbean Spiny Lobster Fisheries   121
             Ernesto A. Chávez

About the Editor   149

Index   151

Related Nova Publications   157

PREFACE

According to the *Oxford English Dictionary* (2018), the word "lobster" originated before the 12th Century C.E. from the Middle English "*lopster*" (Old English "*loppestre*"), meaning "spidery creature" ("*loppe*" = spider). Its earliest known usage in English referred specifically to clawed lobsters in Europe (i.e., *Homarus* spp.). In many contexts, this is still the dominant sense in which the word is used today. However, its usage has also been greatly broadened over the last century to include a more diverse array of marine decapod crustaceans. These include not only the clawed nephropid [Astacidea: Nephropidae (including the homarid lobsters and others)] and reef lobsters (Astacidea: Enoplometopidae), but also blind lobsters (Polychelida), the 'living fossil' glypheid lobsters (Glypheidea), the clawless spiny, slipper, and furry lobsters (Achelata: Palinuridae, Scyllaridae, and 'Synaxidae', respectively), and sometimes the ghost or mud lobsters or 'lobster shrimps' ('Thalassinidea' = Axiidae and Gebiidae); certain anomuran crabs are also commonly referred to as 'squat lobsters' (certain members of Anomura: Chirostyloidea and Galatheoidea). Lobsters possess a relatively elongated body shape with a well-develop muscular abdomen or tail that can be used in swimming escape responses and is typically sought out by fishing industries for human consumption. This distinguishes them from all other reptant or natantian pleocyematan decapod crustaceans, which are instead referred to as 'crabs' or 'shrimps', respectively. Freshwater crayfishes are closely related to the clawed lobsters in the infraorder Astacidea, to which they also belong, but differ from them and all lobsters in their freshwater habitat and lack of a free-living larval phase in the life cycle, and are thus not usually considered to be lobsters.

Many lobsters support important fisheries, and they all play important ecological roles as large, benthic consumers in marine communities; burrowing activities by many groups of lobsters also have important impacts on habitat structuring and sediment bioturbation. The nephropid species *Homarus gammarus* (L., 1758), *Homarus americanus* H. Milne Edwards, 1837, and *Nephrops norvegicus* (L., 1758) support particularly large capture fisheries in eastern North America and along the Atlantic and Mediterranean coasts of Europe, while several species of clawless spiny [*Palinurus* spp., *Panulirus* spp., *Jasus* spp., and *Sagmariasus verreauxi* (H. Milne-Edwards, 1851)] and slipper lobsters also support large and growing fisheries throughout the world. Some species of 'squat lobsters' are also targeted by relatively large fisheries. In much of the world, recent collapses of large finfish stocks due to fisheries overexploitation have led to an increased reliance on invertebrate-targeted fisheries, including lobster fisheries. This is especially apparent in eastern North America, where many coastal communities' economies now rely almost exclusively on the *H. americanus* fishery! This has created a precarious situation, as if lobster stocks were to collapse due to overfishing like those of finfishes before them, then the economic and social impacts on human communities would be considerable, to say nothing of the biological and ecological impacts of such a change on natural marine communities.

Given the above, there is thus much incentive for scientists to study lobsters, especially those species that are commercially exploited. Information on basic species' biology is important for predicting the possible impacts of fishing practices on the stability of their populations, yet such information if often lacking. For example, information on the impacts of temperature changes on lobsters is important in light of future climate change, which in combination with fishing pressure is likely to have complex and important impacts on lobster populations; this is particularly true for larvae, since this phase of the life cycle supplies new recruits to lobster populations, and its ability to disperse in the water column potentially impacts population dynamics and structure. The habitats and behaviors of juvenile lobsters are also important, again because the abundance of juveniles, and in particular survival through the juvenile phase, is an important component of the recruitment of new lobsters to fished populations. Knowing the habitats occupied by juveniles is important so that their abundances can be monitored to predict future changes in numbers of fisheries recruits. If the availability of juvenile habitat is limited, this can limit potential fisheries recruitment. Lobster behaviors can also influence their dispersal among populations (as larvae and adults) and their habitat selection and use and predator avoidance as juveniles, and are thus important components of their biology and ecology that need to be understood. Once information has been collected on a lobster's biology throughout its lifecycle, this can then be used to forecast the potential impacts of different fishing practices (catch limits, minimum legal size, fishing effort, etc.) on exploited lobster populations. Such information can then be used to inform fisheries management and adjust fishing regulations as needed to ensure the sustainability of the industry and the resource on which it depends.

The chapters included in this volume describe studies of the topics listed above, including the thermal biology of *H. americanus* larvae (Chapter 1), the habitats and behaviors of juvenile *H. gammarus* (Chapter 2), and the potential impacts of different approaches to fisheries management on the catches and socioeconomic value of Caribbean spiny lobster [*Panulirus argus* (Latreille, 1804)] fisheries.

Chapter 1 – The development rate of arthropod larvae is related to the performance of rate-controlling enzymes, which generally increases along with development rate with increasing temperature above a lower threshold temperature (T_L), is maximized at an intrinsic optimum temperature (T_Φ), and declines rapidly as an upper threshold temperature is approached (T_H). Knowing these thermal parameters of a species' larvae is important to predict the impacts of thermal variability and future climate change on its recruitment. The Sharpe-Schoolfield-Ikemoto (SSI) model allows these thermal parameters to be estimated and compared using bootstrap 95% confidence intervals. In this chapter, data from eight previous studies of American lobster (*H. americanus*) larvae were used to estimate the values of the T_L, T_Φ, and T_H parameters of the SSI model for lobster larvae and compared them among larval stages (I-IV) and studies. The estimates obtained varied widely and significantly among studies, with there being no consistent and few significant differences in thermal parameters among stages. Most T_Φ estimates were biologically realistic, but the majority of T_L and T_H estimates were not. When all studies' datasets were combined in the same analyses, the T_Φ, T_L, and T_H for American lobster larval development were estimated to be 16.534-17.326°C, 0.095-7.804°C, and 23.483-25.661°C, respectively, and the T_Φ of stage I was significantly lower than those of stages II and III. These results signaled that the changes in coastal summer water temperatures caused by recent and future climate change could initially improve lobster larval development by increasing the exposure of larvae to optimal temperatures for their development, and thus potentially improve larval survival and subsequent recruitment to benthic lobster populations and fisheries; however, as temperatures continue to rise there is the risk that larvae will begin to experience temperatures approaching their T_H, leading to potentially negative effects. However, these results also highlighted the insufficiency of the data collected in previous studies of this species so far to estimate thermal parameters of the SSI model, meaning that new studies rearing larvae over a wider range of more temperatures ae needed to accurately estimate the limiting and optimal temperatures for this species' larval development.

Chapter 2 – Although the European lobster (*H. gammarus*) is in many regards very similar to the closely related American species (*H. americanus*) and supports large fisheries, the biotope (i.e., habitat) of the early benthic juvenile stages of its life cycle is unknown. Whereas the American species' juveniles inhabit shallow coastal cobble grounds in which they can shelter, searches for the European species in similar habitats have thus far been unsuccessful, and aside from a handful of anecdotal observations small, juvenile European lobsters have never been observed in nature. This represents a critical gap in our knowledge of this species' life history, which potentially limits the ability of future changes to its populations and fisheries to be predicted. This chapter used an inferential approach to attempt to estimate the likely biotope of juvenile European lobsters. Information on the morphological, life history, biotope, and behavioral traits (among others) of the adults and juveniles of a selection of 11 benthic decapod species, including *H. gammarus*, *H. americanus*, and 9 other lobsters, crabs, and shrimps inhabiting the same or similar biotopes as adult *H. gammarus*, was used to make inferences about where small juvenile *H. gammarus* may live in nature. Three hypotheses for the reasons these life stage have not been found were also considered, which were that they are: (1) located too deep to be found in shoreline-based field studies; (2) distributed too scarcely to be found; or (3) living deep inside shelters, which are unapproachable by competitors, predators, and humans. Overall, the traits of *H. gammarus* and *H. americanus* were the most similar in these analyses, but with critical differences in the diversity of predators (e.g., finfishes) and decapod competitors in their biotopes, with these being much higher in European than in North American waters, which should have strong impacts on the survival probabilities of settling lobsters and likely require more antipredator behaviors in the European species' juveniles. Larvae and adults of *H. gammarus* are found at much lower densities than those of *H. americanus*, although early-stage larvae of both species are regularly caught in light traps and plankton nets, and field and laboratory studies indicate limited dispersal of larvae in *H. gammarus*, suggesting that its juveniles are not particularly scarce in nature. Both species' juveniles prefer to burrow under rocks, cobble, or other forms of shelter in the laboratory, and *H. gammarus* juveniles exhibit sheltering responses to water drainage in the laboratory suggesting that they may be at least partly adapted for intertidal life. These results appeared to support hypothesis (3) concerning the apparent scarcity of *H. gammarus* juveniles in nature, and refute the others. It was thus inferred that small *H. gammarus* juveniles should be found in shallow, coastal areas, but that they may shelter more deeply within the substrate and emerge less as a means of avoiding the greater diversity of predators and competitors in their native biotope, making them difficult to find. These findings could inform future field surveys to first find, and then monitor the abundances of, this life stage in nature.

Chapter 3 – The Caribbean spiny lobster (*P. argus*) supports fisheries in 25 different countries in the Caribbean region, but harvesting practices vary considerably across this region. Given the high potential for connectivity among these heterogeneous fisheries, this leads to uncertainty in predicting future changes in lobster catches and the potential for stock collapses, which would have important socioeconomic consequences for the communities and nations that depend on *P. argus* fisheries. In this chapter, model simulations using available 15-year sets of historical fisheries data (catches, costs, effort, etc.) were used to estimate the potential impacts of different levels of fishing mortality (F) and different ages of lobsters at first catch (t_c) on the catch, maximum sustainable yield (*MSY*), maximum economic yield (*MEY*), and socioeconomic value (i.e., profit, direct provisioning of jobs, and profits per fisher) of the five main Caribbean lobster fisheries (those in the Bahamas, Brazil, Cuba, Nicaragua, and the United States of America), as well as the entire Caribbean region. The fisheries yield increased with the t_c, and in three cases the yields at the F corresponding to the *MSY* were higher than those at the F required to reach the *MEY*. The profits were higher at higher t_c values in three fisheries, meaning that they were more profitable if harvested at the level of F that attained their *MEY*. One fishery (that in the Bahamas) was not profitable if harvested at the *MSY* level at any age. The social value of most of the fisheries, calculated as the profits per fisher, was the highest at a t_c of 3 years, and again was higher if the F applied was at the *MEY*. However, in the Bahamas' fishery the social value when the F was applied at the *MSY* was negative at any t_c. These results highlighted the differences among the various spiny lobster fisheries in the Caribbean, including in the optimum harvesting strategies that should be applied to them to maximize their catches, sustainability, and socioeconomic value, which has implications to their management. The creation of a multinational organization to guide or enact the management of these fisheries in different countries was recommended to ensure their optimal and sustainable management. This interdisciplinary approach illustrates the importance of balancing biological/ecological and human/socioeconomic needs in the management of lobster fisheries to enable both to be sustained.

In: Lobsters
Editor: Brady K. Quinn

Chapter 1

ESTIMATING INTRINSIC OPTIMUM TEMPERATURES AND LOWER AND UPPER THERMAL THRESHOLDS FOR THE DEVELOPMENT OF AMERICAN LOBSTER LARVAE USING A THERMODYNAMIC MODEL

*Brady K. Quinn**

Department of Biological Sciences, University of New Brunswick,
Saint John, NB, Canada

This work is dedicated to Nan Yao, a precious friend and lobster colleague, whom I miss dearly.

---

* Corresponding Author's E-mail: bk.quinn@unb.ca.

ABSTRACT

Temperature strongly affects the development of American lobster (*Homarus americanus* H. Milne Edwards, 1837) larvae. At temperatures below and above the species' lower (T_L) and upper (T_H) developmental thresholds, respectively, successful development should be prevented, while at temperatures between these warmer conditions should result in faster development, except as the upper threshold is approached. Likewise, the species should have an intrinsic optimum temperature for development (T_Φ), at which its larval development rate is high while its other physiological characteristics are in good condition. These thermal parameters and the temperature-dependent development relationships they define mediate the effects of temperature on recruitment to lobster populations and fisheries, so it is important to know them. However, no previous study has defined all three thermal parameters for the development of American lobster larvae. In this chapter, a thermodynamic model, the Sharpe–Schoolfield–Ikemoto (SSI) model, was fit to development data for American lobster larvae through different stages and at different temperatures from eight previously published studies. Bootstrap 95% confidence intervals (C.I.s) were also generated for SSI thermal parameter (T_Φ, T_L, and T_H) estimates to allow them to be compared among larval stages and source studies. The estimates obtained varied widely and significantly among studies, with there being no consistent and few significant differences in thermal parameters among stages. Most T_Φ estimates were biologically realistic, but the majority of T_L and T_H estimates were not. When all studies' datasets were combined in the same analyses, the T_Φ, T_L, and T_H for American lobster larval development were estimated to be 16.534-17.326°C, 0.095-7.804°C, and 23.483-25.661°C, respectively. The T_Φ of stage I was significantly lower than those of stages II and III, but all other differences in thermal parameters among stages were non-significant. Further, while all T_Φ and most T_L estimates were considered realistic, all T_H estimates were slightly too low. The SSI model was derived based on enzyme thermodynamics, and thus should provide useful information concerning lobster larval developmental optima and limits. The results found suggest that as ocean temperatures experienced by lobster larvae in nature more often exceed the T_Φ for their development due to climate change, larval development and survival may be impaired, which will have important impacts on lobster recruitment. However, none of the previous studies analyzed were conducted over a wide enough thermal range to allow a conclusive fit of their data to be done with the SSI model. Therefore, a future rearing study conducted over a wide thermal range, followed by analyses with the SSI model, is needed to better define the limiting and optimum temperatures for the development of this species' larvae.

Keywords: American lobster, larva, development, threshold temperature, intrinsic optimum temperature, thermodynamics, Sharpe–Schoolfield– Ikemoto (SSI) model, bootstrap confidence intervals

1. INTRODUCTION

Environmental temperatures have important and powerful impacts on the physiology and ecology of ectothermic animals (Bělehrádek 1935; Anger 2001; Angilletta Jr. 2006; Byrne 2011; Jost et al. 2012; Shi et al. 2011, 2012a, b, 2013, 2017), including lobsters (McLeese 1956; Boudreau et al. 2015; Quinn 2017a). Larval development is a particularly important temperature-sensitive process in the lifecycles of ectotherms, as it must be successfully completed for there to be recruitment to later life stages and for populations to be maintained (Schmalenbach and Franke 2010; Vaughn and Allen 2010; Byrne 2011; Pineda and Reyns 2018). At temperatures that are too low or too high, development cannot be completed; these limiting temperatures are termed the lower (T_L) and upper developmental thresholds (T_H), respectively, and vary among species (Jost et al. 2012; Shi et al. 2017) and perhaps developmental stages of the same species (Campbell et al. 1974; Brière et al. 1999; Quinn 2017a). The rate at which larval development proceeds is also important, as a longer larval developmental period results in greater exposure of larvae to predation risks, environmental stress, possible dispersal away from juvenile habitat, and other sources of mortality, and thus decreases the probability of them surviving the larval period and recruiting to the adult population (Pechenik 1999; Vaughn and Allen 2010; O'Connor et al. 2007; Pineda and Reyns 2018; Gendron et al. in press). Over the majority of temperatures likely to be experienced by a species between its lower and upper thermal thresholds, higher temperatures result in faster development and thus shortened larval durations (Angilletta Jr. 2006; Campbell et al. 1974; Ikemoto and Takai 2000), and the relationship between temperature and development rate is usually linear across much of this range, although near extreme values it becomes nonlinear (Brière et al. 1999; Jost et al. 2012;

Shi et al. 2011; 2012a,b; 2013; 2017). However, as temperatures approach the upper threshold this beneficial effect of temperature breaks down and development speeds up less and less, and then eventually slows rapidly just before the threshold is reached (Schoolfield et al. 1981; Yamamoto et al. 2017). In light of ongoing and future climate change, there is much interest and importance in obtaining a better understanding of the effects of temperature changes on various organisms (Byrne 2011; Caputi et al. 2013; Pinsky et al. 2013). It is particularly important that the thermal thresholds and temperature-dependent development rates of ecologically and/or economically important species are quantified. However, for many species such information has not yet been sufficiently assessed due to the high costs and methodological challenges of needing to rear larvae over a wide range of temperatures to do so (e.g., Quinn 2017b; but see Yamamoto et al. 2017).

The American lobster, *Homarus americanus* H. Milne Edwards, 1837 (Crustacea: Decapoda: Astacidea: Nephropidae), is a large, clawed species of lobster than inhabits the coastal marine benthos of the Atlantic Shelf of eastern North America, from Cape Hatteras, NC, USA to Labrador, NL, Canada (Lawton and Lavalli 1995). This species is presumed to be an important heterotrophic component of many marine communities, and supports the most lucrative and socioeconomically important fishery in the region spanning Cape Cod, MA, USA to Newfoundland, Canada (Wahle et al. 2013). During the spring-summer months (ca. May-October), lobster larvae are released by benthic females, and then spend from 2-8 weeks or more (depending on water temperature) in the water column, living as planktonic and/or pelagic raptorial predators (Phillips and Sastry 1980; Ennis 1995). Lobster larvae develop through three zoea or mysis stages, termed larval stages I, II, and III, which are followed by a single decapodid stage, termed stage IV (and commonly called a 'postlarva') (Hadley 1906; MacKenzie 1988; Ennis 1995); stage IV lobsters are strong swimmers that settle to the benthos when they encounter suitable substrate, after which they moult into the first juvenile instar (stage V) (Lawton and Lavalli 1995). Some strong correlations have been found between the supply of potential settlers (i.e., larvae) to a given locality and its fisheries recruitment (e.g., Miller 1997; Wahle et al. 2004; Quinn et al. 2017; Gendron et al. in press), so it is likely that any factor that impacts larval survival and/or planktonic dispersal patterns (which affect the spatial distribution of settlers) could strongly impact lobster populations and the fisheries they support (Caputi et al. 2013; Boudreau et al. 2015; Jaini et al. 2018). Changing water temperature due to climate change is one such factor, as such change has the potential to expose lobster larvae to temperatures exceeding their developmental thresholds and/or to temperatures that lengthen larval duration (Quinn and Rochette 2015; Quinn 2017a). Therefore, it is clearly important that we know the thermal thresholds and temperature-dependent development rates of American lobster larvae to allow such changes and their potential impacts to be predicted.

A number of studies have investigated the impacts of temperatures on American lobster larval development times (e.g., Hadley 1906; Templeman 1936; Hughes and Matthiessen 1962; Ford et al. 1979; MacKenzie 1988; Hudon and Fradette 1988; Annis et al. 2007; Quinn et al. 2013; Waller et al. 2017; Harrington et al. 2019). Although studies have been done at rearing temperatures within the range of 6.7 to 26.3°C, no single study has been done over this full range, and none have yet investigated the larval development of this species across its full range of potentially biologically relevant temperatures (ca. 0-30°C or higher: Quinn and Rochette 2015). Thus, we do not yet have a detailed picture of its larval development curve, including the point at which rising temperatures cease to increase development rates (Quinn 2017a); although it should be noted that based on results of Templeman (1936) and Ford et al. (1979), development rates may plateau between 19.2 and 23.8°C and begin to decline slightly between 24.2 and 26.3°C. Further, although a handful of studies have attempted to estimate lethal temperature limits of lobster larvae (0°C and 28-35.5°C: Huntsman 1924; Sastry and Vargo 1977; Gruffydd et al. 1979), no comprehensive investigation of the thermal limits of larval development (i.e., using long-term exposures of large numbers of larvae to extreme temperatures) has yet been done (Quinn 2017a).

Until such studies are done, the only means available to assess potential developmental threshold temperatures for lobster larvae is the application of development functions that can estimate such thresholds to data from previous studies (Quinn 2017a, b). For example, Quinn (2017b) applied the linear sum (Winberg 1971), linear rate (Campbell et al. 1974), Bělehrádek (Bělehrádek 1935; McLaren 1963), modified Arrhenius (Guerrero et al. 1994), and Brière-2 (Brière et al. 1999) functions to data from MacKenzie (1988) and Quinn et al. (2013) to estimate T_L and/or T_H thresholds of lobster larvae of 0-8.8°C and 22.5-41.4°C, respectively (Quinn 2017a). However, many of these estimates were not considered biologically realistic and/or were made with limited power (see Quinn 2017b for discussion), and could further not be statistically compared among larval stages or data sources. These functions also did not allow the optimum temperature for larval development to be estimated (see below). There thus remains the potential for better estimates of developmental threshold temperatures to be made for this species.

A prominent and potentially very useful (but also extremely complex) development function that was not examined or applied to lobster data by Quinn (2017b) was the Sharpe–Schoolfield–Ikemoto (SSI) model (Schoolfield et al. 1981; Shi et al. 2011; Ikemoto et al. 2013). This model was derived from studies of enzyme thermodynamics, and equates the limiting temperatures for development (T_L and T_H) to those at which a hypothetical enzyme (presumably one controlling and/or affecting development) has a 50% probability of being in either an active or temperature-inactivated state (Ikemoto et al. 2013). It thus has the potential to produce more realistic and meaningful estimates of thermal limits than many of the other models applied previously to lobsters. This model has been applied extensively in studies of insects and arachnids (e.g., Ikemoto 2005, 2008; Shi et al. 2011, 2012a, b, 2013, 2017; Jafari et al. 2012; Ikemoto and Egami 2013; Padmavathi et al. 2013; Sreedevi et al. 2013; Quinn 2017b), and has recently been applied to a few crustaceans (Yamamoto et al. 2017). The SSI model is represented by the equation:

$$d(T) = \frac{\rho_\Phi \frac{T}{T_\Phi} exp\left[\frac{\Delta H_A}{R}\left(\frac{1}{T_\Phi}-\frac{1}{T}\right)\right]}{1++exp\left[\frac{\Delta H_L}{R}\left(\frac{1}{T_L}-\frac{1}{T}\right)\right]+exp\left[\frac{\Delta H_H}{R}\left(\frac{1}{T_H}-\frac{1}{T}\right)\right]} \qquad (1)$$

where $d(T)$ is the development rate (1/development time, in 1/day) at a given absolute temperature, T (in degrees Kelvin (K), such that 0°C = 273.15 K), R is the gas constant (1.987 cal/deg/mol), ΔH_A is the enthalpy (in cal/mol) of a reaction that is catalyzed by a hypothetical enzyme, T_L and T_H are the lower and upper threshold temperatures (in K) at which the enzyme is half active and either half low-temperature inactivated or half high-temperature inactivated, respectively, ΔH_L and ΔH_H are the changes in enthalpy (in cal/mol) associated with the low- or high-temperature inactivation of the enzyme, respectively, T_Φ is the intrinsic optimum temperature (in K) of the reaction at which the probability of the enzyme being in an active state is maximized, and ρ_Φ is the development rate (in 1/day) at the T_Φ assuming no enzyme inactivation (Sharpe and DeMichele 1977; Schoolfield et al. 1981; Shi et al. 2011; Ikemoto et al. 2013).

The intrinsic optimum temperature, T_Φ, is related to the other parameters of the SSI model, and thus may be the most important and informative thermal parameter to estimate (Ikemoto et al. 2013). In previous work, the T_Φ was assumed to be equivalent to 25°C (in the SS model of Sharpe & DeMichele (1977)), but more recent studies using the SSI model have permitted the value of the T_Φ to vary as appropriate for the species and dataset under consideration (Shi et al. 2011, 2012a, 2013, 2019; Ikemoto et al. 2013). Traditionally, the 'optimum' temperature for the larval development of a species was considered to be the temperature at which its development rate was maximized (e.g., Amarasekare and Savage 2012). However, this is not really the case, as an increasing body of work has demonstrated that at temperatures causing the most rapid development possible a number of other physiological processes are impaired, causing reduced growth, survival, etc. (Ikemoto 2008; Martin and Huey 2008; Forster et al. 2011; Corkrey et al. 2012; Shi et al. 2012a, 2013, 2019); indeed, for lobsters this seems to be the case as well (Quinn 2017a; Harrington et al. 2019). Thus, the true intrinsic optimum temperature for the development of a species should be below the one at which development rate is maximized (Martin and Huey 2008; Shi et al. 2019), and algorithms used to estimate SSI model parameters should not constrain the value of T_Φ to be equal to either 25°C or the temperature at which the development rate is maximized.

There is obviously interest in determining the intrinsic optimum temperature for the development of a species under consideration. Many species of insects and crustaceans, for example, are reared in laboratory or hatchery settings for use as food for humans or livestock or to supplement natural stocks of their species (Van Olst et al. 1976; Carlberg and Van Olst 1977; Ford et al. 1979; Quinn 2017b), and to maximize their production and efficiency such efforts should be carried out at temperatures that maximize larval development rates while avoiding negative thermal effects on survivorship and other characteristics. It is also useful to know the optimum temperature for a species when making predictions about its dynamics in nature, as temperatures below and above the optimum can have important negative impacts on larval survival and supplies of recruits (Quinn and Rochette 2015; Quinn 2017a) and threshold temperatures may determine the boundaries of geographic distributions (Boudreau et al. 2015). There is a long history of studies of American lobster, which does suggest that the optimum temperature for larvae of this species is somewhere around 18-20°C (e.g., Van Olst et al. 1976; Carlberg and Van Olst 1977; Ford et al. 1979; Bartley et al. 1980; Quinn 2017a); however, such estimates vary among studies and have not been made while considering the thermodynamics of the enzymes controlling development. Therefore, it may be useful to estimate the intrinsic optimum temperatures for American lobster larval development using the SSI model.

Recently, techniques have also been developed that allow the thermal parameters T_Φ, T_L, and T_H estimated by the SSI model to be compared among datasets based on analyses of bootstrap confidence intervals (Ikemoto et al. 2013). The possibility that such thermal characteristics differ among developmental stages within the same species has previously been raised (Anger 2001; Schmalenbach and Franke 2010; Jafari et al. 2012), and has implications to changes in the timing of events in nature under climate change; for example, if a certain stage is more sensitive to elevated temperatures than others, then it may represent a bottleneck for recruitment to all later life stages (Gruffydd et al. 1979; Caputi et al. 2013; Quinn 2019). Being able to compare limiting and optimum temperatures for the development of American lobster larvae among stages, and perhaps also among different source populations (e.g., to test for local adaptation; Quinn et al. 2013) would also be highly useful.

Therefore, in this chapter the SSI development model was fit to data for the development of American lobster larvae at different temperatures obtained from the literature using a series of statistical packages and techniques developed by Ikemoto et al. (2013). This was done to generate new and hopefully improved estimates of the lower and upper threshold temperatures for American lobster larval development compared to those produced in previous studies, and to compare these estimates statistically among larval stages and different source studies. During this procedure, estimates of the intrinsic optimum temperature for the larval development of this species were also produced and compared, which has not been done before. The results obtained were discussed in light of previous research on this species, as well as in relation to the species' larval ecology and potential impacts of climate change. The quality of the estimates obtained was also critically evaluated to make recommendations for future studies of this species' temperature-dependent larval development.

2. MATERIALS AND METHODS

2.1. Sources of Development Data

Development times of American lobster larvae at different temperatures were obtained from 8 previously published studies (Table 1). These studies reported development times of American lobster larvae at three or more (up to 25) different temperatures, which were as low as 6.7°C, as high as 26.3°C, and covered a thermal range of 3.2 to 17.1°C (mean: 8.45°C) in total (Table 1). Many additional studies besides these 8 have also examined American lobster larval development times at different temperatures (e.g., Annis et al. 2007; Waller et al. 2017) but only compared two temperatures, meaning their data could not be fit with a complex curve like the SSI model; these other studies' data were thus not used herein.

The studies whose data were used were carried out in either a laboratory or hatchery setting (6/8 studies) or in the field (2 studies), and used larvae that originated from different locations across the species' range, as reviewed by Quinn et al. (2013) (Table 1). Larvae were reared individually (5 studies) or communally (2 studies), or sampled in field plankton tows (1 study) (Table 1). One field study (Hadley 1906) did not rely on plankton tows, but rather held groups of larvae in scrim bags moored in a harbor. Rearing temperatures were controlled and held ~constant in 4 studies, while in 3 other studies temperatures were allowed to undergo natural fluctuations, and 1 study tested both controlled and fluctuating temperatures (Table 1). All 8 studies reported total development times from hatch to the moult to stage IV (i.e., the combined duration of stages I, II, and III), and most (5/8 studies) also reported individual development times through each of stages I, II, and III (Table 1). However, development times for stage IV, and therefore total development times from hatch to the moult to stage V (i.e., the combined duration of stages I, II, III, and IV), were only reported by 3 of these studies (Table 1).

For each study, the mean development time (in days) spent in each stage (I, II, III, and IV) or combination of stages (I-III combined and I-IV combined) at each studied temperature was either taken directly from the published paper (if reported) or calculated from the data presented therein. These mean temperature-dependent development times were then used in all subsequent analyses.

Table 1. Details of American lobster larval rearing studies from which data were obtained

Study	Stages	Temperatures (°C) (range, # tested)	Source population	Rearing conditions	Temperature control
Hadley (1906)	I-III	15.6-22.2 (6.6, 3)	Rhode Island, USA	Communal (field)	Variable
Templeman (1936)	I, II, III, I-III	6.7-23.8 (17.1, 11)	Bay of Fundy and Northumberland Strait, Canada	Individual (lab)	Constant
	IV, I-IV	9.1-19.2 (10.1, 9)			
Hughes and Matthiessen (1962)	I-III	14.0-22.3 (8.3, 25)	Massachusetts, USA	Communal (hatchery)	Variable
Ford et al. (1979)	I-III	16.9-26.3 (9.4, 5)	Hatchery, origin not specified (study done in California, USA)	Individual (lab)	Constant and variable (both tested)
MacKenzie (1988)	I, II, III, I-III, IV, I-IV	9.8-22.0 (12.2, 5)	Southwest Nova Scotia and Northumberland Strait, Canada	Individual (lab)	Constant
Hudon and Fradette (1988)	I	11.3-17.5 (6.2, 12)	Magdalen Islands, QC, Canada	Field sampling	Variable
	II	12.3-17.5 (5.2, 11)			
	III, I-III	14.3-17.5 (3.2, 10)			
	IV, I-IV	14.3-17.5 (3.2, 9)			
Quinn et al. (2013)	I, II, III, I-III	10.3-22.2 (11.9, 4)	Gaspé-Nord, QC, Canada	Individual (lab)	Constant
Harrington et al. (2019)	I, II, III, I-III	14.0-22.0 (8.0, 4)	Maine, USA	Individual (lab)	Constant

2.2. Estimating SSI model Parameters

The mean development time data for each study and stage or combination of stages at different temperatures were fit with the SSI model using the *OptimSSI* package by Ikemoto et al. (2013) in R v.3.1.1 (R Core

Team 2014). This package estimated the parameters T_Φ, T_L, T_H, ρ_Φ, ΔH_A, ΔH_L, and ΔH_H of the SSI model in equation (1) for each dataset, which were output as T-Phi, TL, TH, rho-Phi, HA, HL, and HH, respectively. It also produced χ^2 and R^2 goodness-of-fit values for the fit of the model to each dataset, which were output as Chi-square and R-square, respectively. This package was used because the SSI model is extremely complex relative to other temperature-dependent development functions and is not always possible to fit to data using all statistical programs' nonlinear regression procedures (Quinn 2017b). Although earlier programs were made to estimate SSI model parameters (Ikemoto 2008; Shi et al. 2011), they tended to be too slow for practical purposes, whereas the package introduced by Ikemoto et al. (2013) is able to estimate parameters very quickly, which is particularly important for its use when calculating bootstrap confidence intervals for these estimates (see section 2.3).

The main inputs to the *OptimSSI* program are temperatures (in °C) and development times (in days); the package then converts the temperatures into degrees Kelvin (K) and the development times into development rates (1/day). The user must also specify what portion of the dataset represents the putative 'linear' portion of the temperature-development curve. Over the specified range, a linear fit is then made of the temperature-development rate relationship (Winberg 1971; Campbell et al. 1974; Ikemoto and Takai 2000), from which initial estimates of the parameters T_L and T_H are made (Ikemoto et al. 2013; Quinn 2017b). These initial estimates are then used as starting values for these parameters in the nonlinear fitting of the data to the SSI model.

For American lobster larvae, very few studies have examined development over a sufficiently broad range of temperatures to provide a good idea of where the 'linear' portion of their temperature-development relationship occurs (Quinn 2017a, b; Table 1). The lower boundary should occur where a decrease in temperature results in a decrease in the development rate that becomes greater in magnitude with each subsequent 1° decrease, while the upper boundary should occur at the point where an increase in temperature ceases to cause an increase in development rate, but rather begins to cause a decrease (Campbell et al. 1974; Jost et al.

2012). Several studies have identified 12°C as a potential lower 'threshold' affecting American lobster larvae, below which survival and settlement decrease markedly and development becomes very slow (MacKenzie 1988; Annis et al. 2013; Quinn 2017a). Meanwhile, the only study that observed slowing development of American lobster larvae with increasing temperature was Ford et al. (1979), who observed slower development at 26.3°C than that at 24.2°C, although Templeman (1936) reported that development time changed very little as the temperature increased from 19.2 to 23.8°C. Therefore, in this chapter the lower boundary of the linear portion of the development curve was set at 12°C and the upper boundary was set at 24.3°C. This was fixed for analyses of all studies and stages due to limited information on whether this should vary among stages or the region of larval origin.

In the calculation of parameter estimates, the *OptimSSI* package initially sets the values of T_L and T_H (in K) to those estimated from the linear fit, T_Φ to 298.15 K (25°C), and ΔHL and ΔHH to −50,000 and 50,000 cal/mol, respectively; for other details of model initialization and parameter estimation algorithms, see Ikemoto et al. (2013). After performing all calculations, the program reports all parameter estimates for a given dataset as its final output. For ease of interpretation, all temperature parameter estimates, although reported in K by the program and used as such within the SSI function, were converted to °C for presentation in all tables and figures herein. Plots of the SSI curves for each stage and study dataset were made in R using the *SSIPlot* function of Ikemoto et al. (2013).

2.3. Calculating Bootstrap Confidence Intervals of Parameter Estimates

Because specialized algorithms are required to estimate parameters of the SSI model, it can be difficult to obtain reasonable estimates of the uncertainty around these parameter estimates, which also means that they cannot always be compared statistically (e.g., among larval stages, studies, populations of origin, etc.). Ikemoto et al. (2013) solved this issue by introducing the use of bootstrapping techniques to estimate 95% confidence intervals (95% C.I.s) of SSI model parameter estimates. Bootstrapping is a family of computerized techniques that perform random resamplings of a dataset or distribution with replacement to derive an estimate of the accuracy or repeatability of some sort of estimate made from a sample (DiCiccio and Efron 1992, 1996; Efron and Tibshirani 1994). Ikemoto et al. (2013) introduced two packages to estimate 95% C.I.s of SSI model parameter estimates using different bootstrap techniques in R. The first of these, *BCaSSI*, provides bootstrapped 95% C.I.s for all parameters estimated by the *OptimSSI* package using both the bootstrap percentile and the bias-corrected and accelerated bootstrap (BC_a) methods (Efron and Tibshirani 1994). The second, *mABCSSI*, provides bootstrapped 95% C.I.s for T_Φ only using the approximate bootstrap confidence intervals (ABC) technique (DiCiccio and Efron 1992, 1996; Efron and Tibshirani 1994), which Ikemoto et al. (2013) identified as the most appropriate approach to use for making comparisons of T_Φ values. Therefore, once *OptimSSI* had generated SSI parameter estimates for a given dataset, these algorithms were used to estimate 95% C.I.s of all parameter estimates in the present chapter's analyses.

2.4. Comparisons of Estimates among Larval Stages and Studies

Comparisons herein were focused on the parameters of the SSI model representing potential thermal thresholds for the development of lobster larvae (T_L and T_H), as well as the estimated intrinsic optimum temperature for larval development (T_Φ). As a first set of preliminary tests, one-way analyses of variance (ANOVAs) were performed in R to coarsely test whether estimated T_L, T_H, and T_Φ values differed among American lobster larval stages, with different studies treated as replicates. Overall mean values of each parameter ± 95% C.I.s were also calculated for each stage or combination of stages across studies and plotted. Whether the results of these tests would lead to different conclusions than those of analyses based on bootstrapped 95% C.I.s (see below) was also considered.

Bootstrapped 95% C.I.s formed the primary basis of comparisons among SSI model parameters among American lobster larval stages and studies. Comparisons among the estimates for a given group (stage or study) were made using analysis of confidence intervals (Cummings et al. 2007). In this approach, if the 95% C.I.s of one group overlap with the mean value of another, then the two groups can be concluded to not be significantly different (p > 0.05), whereas if both groups' 95% C.I.s do not overlap with one another's means (i.e., the 95% C.I. of the difference between the two groups does not overlap with zero; Lin et al. 2018; Shi et al. 2017, 2019) then they are significantly different (p < 0.05) (Cummings et al. 2007). SSI model parameter estimates were compared in place of mean values in the present chapter, and the 95% C.I.s were those produced by bootstrapping for each parameter.

Two series of comparisons were made. First, each of these estimates was compared among different larval stages (I, II, and III, and IV if data were available) for each study that provided data on multiple stages (Table 1). Second, each of these estimates was compared among different studies for each stage (I, III, III, and IV) and combination of stages (I-III and I-IV). T_L and T_H estimates were compared using the 95% C.I.s calculated for them based on both the bootstrap percentile and BC_a methods. T_Φ estimates were also compared using 95% C.I.s generated using these two techniques, in addition to using those generated with the ABC method; however, given the conclusions of Ikemoto et al. (2013) more weight was given to the results of comparison of T_Φ estimates based on ABC-derived 95% C.I.s.

Whether overall and per-study estimates of T_L, T_H, and T_Φ values were biologically realistic was also considered. Since T_L and T_H are potential temperatures at which larval development could be impaired and prevented, estimates of these parameters that fell within the range of temperatures over which successful larval development has been observed in previous studies (6.7–26.3°C; Table 1; Quinn 2017a) were considered unrealistic. Estimates of T_L, T_H, or T_Φ beyond the likely range of temperatures that are biologically relevant to American lobster larvae (e.g., those < 0°C and >> 40°C; Quinn and Rochette 2015; Quinn 2017a, b) were also suspected of being unlikely to be true. If $T_L > T_H$, $T_L > T_\Phi$, or $T_\Phi > T_H$, this was also obviously unrealistic. Previous studies have suggested (based on various criteria) that temperatures of ca. 18-20°C are 'optimal' for American lobster larval growth (Van Olst et al. 1976; Carlberg and Van Olst 1977; Ford et al. 1979; Bartley et al. 1980; Quinn 2017a), so T_Φ was expected to be near this range. A T_Φ estimate that was not excessively high, but high enough to be within the range over which higher temperatures begin to slow down development (i.e., beyond the upper bound of the 'linear' portion of the development curve, which was herein considered to be ca. 24.2-26.3°C (Ford et al. 1979)), was also suspected of being unrealistic.

2.5. Combined Analyses Using All Studies' Data

As only a relatively small sample size and limited range of temperatures was captured in most of the individual studies from which data were extracted herein (Table 1), a further set of analyses was done in which all 8 studies' data were combined into a single dataset. All of the procedures described above (sections 2.2-2.4) were then carried out on the combined dataset for each stage and combination of stages. Whether the use of the combined dataset resulted in different conclusions regarding the thermal parameters of lobster larval development than analyses of individual studies' datasets was then examined.

3. RESULTS

3.1. Overall Findings

Estimates of the parameters of the SSI model when fit to development time data of American lobster larvae from different studies and their bootstrapped 95% C.I.s are presented in Tables 2-8. The corresponding temperature-dependent development rate curves based on the SSI model parameters estimated for each of these datasets are plotted in Figures 1-6.

Table 2. SSI parameter estimates for data from Templeman (1936)

	Parameter	Estimate	LCI (BP/ABC)	UCI (BP/ABC)	LCI (BC$_a$)	UCI (BC$_a$)
Stage I	T-Phi	17.673	12.484	19.371	12.713	21.378
			16.832	18.049		
	TL	5.378	-138.230	8.744	-132.216	8.898
	TH	24.423	20.528	880.038	-33129.340	29.906
	rho-Phi	0.2399	0.1057	0.3048	0.1256	0.4614
	HA	2.175E+04	1.628E+04	3.335E+04	1.548E+04	3.230E+04
	HL	-6.971E+04	-2.277E+05	-9.702E+03	-5.881E+05	-1.922E+04
	HH	1.562E+05	1.154E+04	3.752E+05	6.515E+04	5.542E+06
	Chi-square	3.694E-02	1.332E-04	7.348E-02	2.936E-03	4.634E-01
	R-square	0.9249				
Stage II	T-Phi	16.786	13.459	19.606	14.004	20.187
			16.619	17.018		
	TL	7.936	6.854	9.254	6.809	8.812
	TH	27.868	19.706	37.513	21.294	134.736
	rho-Phi	0.1828	0.1211	0.2346	0.1298	0.2480
	HA	1.747E+04	1.481E+04	2.428E+04	1.310E+04	2.204E+04
	HL	-8.393E+04	-2.086E+05	-7.447E+04	-1.086E+05	-5.281E+04
	HH	6.863E+04	2.842E+04	2.908E+05	7.548E+03	1.526E+05
	Chi-square	2.417E-03	1.152E-04	3.997E-03	7.981E-04	5.045E-03
	R-square	0.9953				
Stage III	T-Phi	18.240	13.937	20.445	15.161	22.233
			17.882	18.561		
	TL	6.896	4.902	9.508	0.725	9.384
	TH	28.662	18.017	53.484	19.066	70.220
	rho-Phi	0.1387	0.0858	0.1697	0.0978	0.1882
	HA	1.729E+04	1.526E+04	2.035E+04	1.430E+04	1.999E+04
	HL	-5.736E+04	-2.139E+05	-4.339E+04	-1.230E+05	-9.680E+03
	HH	7.083E+04	1.659E+04	6.238E+05	6.549E+03	2.639E+05
	Chi-square	5.241E-03	3.053E-04	6.819E-03	3.509E-03	9.160E-03
	R-square	0.9902				
Stage IV	T-Phi	18.005	12.417	19.649	14.217	24.193
			17.111	18.026		
	TL	-1.252	-129.485	9.854	-144.950	9.479
	TH	48.215	15.865	275.492	18.437	1602.177
	rho-Phi	0.0600	0.0327	0.0711	0.0401	0.0837
	HA	1.736E+04	1.114E+04	2.293E+04	1.256E+04	2.620E+04
	HL	-4.538E+04	-8.102E+05	-9.791E+03	-2.452E+05	-1.366E+03
	HH	3.205E+04	8.354E+03	1.376E+06	1.426E+01	1.313E+05
	Chi-square	2.460E-03	7.735E-06	3.596E-03	1.284E-03	4.378E-02
	R-square	0.9601				

Notes: LCI and UCI = lower and upper 95% C.I.s estimated with the bootstrap percentile (BP), BC$_a$, or ABC methods; ABC-derived 95% C.I.s were calculated for T-Phi only, and are presented in the same columns as BP-derived 95% C.I.s, but one row beneath them; aE±b = a × 10$^{±b}$; N/A = could not be estimated; NaN = non-numeric estimate; T-Phi, TL, and TH estimates are presented in units of °C; for the units and meanings of all SSI model parameters, see equation (1) in the Introduction, and for their abbreviations in this Table and in *OptimSSI* output, see section 2.2.

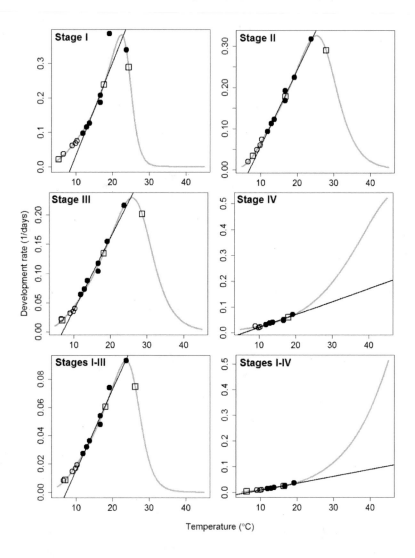

Figure 1. SSI model curves fit to development rates (y-axes, in 1/days) of American lobster larvae through different stages (different panels) at different temperatures (x-axes, in °C), using data obtained from Templeman (1936). Solid circles = observed data within the linear portion of the development curve; open circles = observed data outside the linear range; gray lines = the SSI development curve fit to all observed data; black lines = linear fit of data in the linear portion of the development curve; open squares = estimates of (from left to right) T_L, T_Φ, and T_H based on the plotted data and curves. All plots made in R with the *SSIPlot* function of Ikemoto et al. (2013).

Table 3. SSI parameter estimates for data from MacKenzie (1988)

	Parameter	Estimate	LCI (BP/ABC)	UCI (BP/ABC)	LCI (BC$_a$)	UCI (BC$_a$)
Stage I	T-Phi	15.381	12.810	18.215	12.743	18.156
			15.362	15.928		
	TL	7.764	-4.725	9.961	-4.725	9.961
	TH	29.231	17.582	41.979	23.014	43.637
	rho-Phi	0.1947	0.1359	0.2587	0.1352	0.2586
	HA	1.873E+04	1.579E+04	2.040E+04	1.579E+04	2.040E+04
	HL	-9.471E+04	-2.358E+05	-2.325E+04	-2.105E+05	-1.771E+04
	HH	4.741E+04	2.068E+04	2.292E+05	1.932E+04	1.664E+05
	Chi-square	6.853E-05	1.698E-11	6.853E-05	1.966E-05	6.853E-05
	R-square	0.9997				
Stage II	T-Phi	17.590	12.540	19.208	15.508	19.863
			17.357	18.082		
	TL	8.101	-1.960	13.975	-2.538	9.476
	TH	23.374	17.217	49.415	16.545	24.243
	rho-Phi	0.2261	0.1024	0.2740	0.1671	0.2896
	HA	2.305E+04	1.403E+04	2.744E+04	1.403E+04	2.744E+04
	HL	-8.526E+04	-6.527E+05	-4.423E+04	-5.950E+05	-4.423E+04
	HH	1.675E+05	1.688E+04	9.039E+05	3.133E+04	1.217E+06
	Chi-square	3.457E-04	2.724E-11	9.899E-04	1.471E-09	1.040E-03
	R-square	0.9987				
Stage III	T-Phi	15.056	13.031	19.416	12.047	16.884
			15.033	20.345		
	TL	8.991	5.357	10.995	6.548	11.020
	TH	29.734	19.283	59.568	19.959	61.541
	rho-Phi	0.1167	0.0808	0.1916	0.0649	0.1531
	HA	2.305E+04	1.933E+04	3.129E+04	1.933E+04	3.129E+04
	HL	-1.103E+05	-3.791E+05	-5.046E+04	-4.074E+05	-6.158E+04
	HH	3.556E+04	1.908E+04	2.341E+05	1.583E+04	1.122E+05
	Chi-square	2.429E-04	2.288E-11	2.429E-04	8.535E-05	2.589E-04
	R-square	0.9990				
Stage IV	T-Phi	14.363	12.868	23.103	12.536	15.902
			N/A	N/A		
	TL	9.258	-5.514	10.912	0.647	11.147
	TH	42.317	17.054	80.326	23.999	123.036
	rho-Phi	0.0403	0.0343	0.0763	0.0336	0.0448
	HA	1.271E+04	7.508E+03	1.736E+04	7.508E+03	1.736E+04
	HL	-2.072E+05	-6.369E+05	-6.927E+04	-7.715E+05	-7.658E+04
	HH	2.975E+04	1.713E+04	1.540E+06	1.159E+04	9.656E+04
	Chi-square	2.394E-04	3.186E-12	3.422E-04	3.250E-08	3.775E-04
	R-square	0.9927				

Notes: LCI and UCI = lower and upper 95% C.I.s estimated with the bootstrap percentile (BP), BC$_a$, or ABC methods; ABC-derived 95% C.I.s were calculated for T-Phi only, and are presented in the same columns as BP-derived 95% C.I.s, but one row beneath them; aE±b = a × 10$^{\pm b}$; N/A = could not be estimated; NaN = non-numeric estimate; T-Phi, TL, and TH estimates are presented in units of °C; for the units and meanings of all SSI model parameters, see equation (1) in the Introduction, and for their abbreviations in this Table and in *OptimSSI* output, see section 2.2.

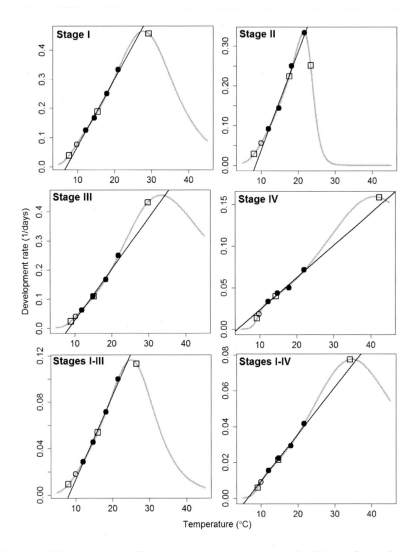

Figure 2. SSI model curves fit to development rates (y-axes, in 1/days) of American lobster larvae through different stages (different panels) at different temperatures (x-axes, in °C), using data obtained from MacKenzie (1988). Solid circles = observed data within the linear portion of the development curve; open circles = observed data outside the linear range; gray lines = the SSI development curve fit to all observed data; black lines = linear fit of data in the linear portion of the development curve; open squares = estimates of (from left to right) T_L, T_Φ, and T_H based on the plotted data and curves. All plots made in R with the *SSIPlot* function of Ikemoto et al. (2013).

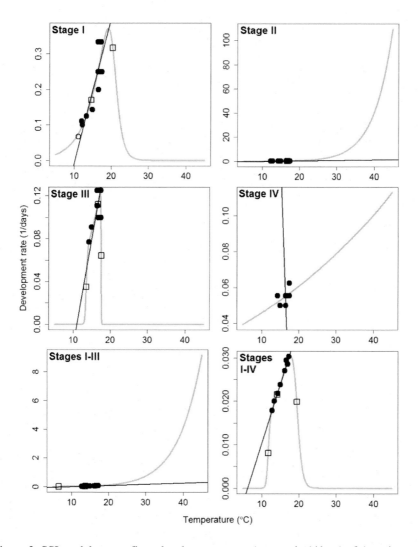

Figure 3. SSI model curves fit to development rates (y-axes, in 1/days) of American lobster larvae through different stages (different panels) at different temperatures (x-axes, in °C), using data obtained from Hudon and Fradette (1988). Solid circles = observed data within the linear portion of the development curve; open circles = observed data outside the linear range; gray lines = the SSI development curve fit to all observed data; black lines = linear fit of data in the linear portion of the development curve; open squares = estimates of (from left to right) T_L, T_ϕ, and T_H based on the plotted data and curves. All plots made in R with the *SSIPlot* function of Ikemoto et al. (2013).

Brady K. Quinn

Table 4. SSI parameter estimates for Hudon and Fradette (1988)

	Parameter	Estimate	LCI (BP/ABC)	UCI (BP/ABC)	LCI (BC$_a$)	UCI (BC$_a$)
Stage I	T-Phi	14.715	12.863	16.957	12.805	16.908
			14.653	17.505		
	TL	-37.113	-117.530	11.704	-253.580	-21.934
	TH	20.394	17.528	439.689	-2705.501	89.515
	rho-Phi	0.1819	0.1132	0.2912	0.1124	0.2888
	HA	3.707E+04	3.005E+04	5.521E+04	2.923E+04	4.857E+04
	HL	-7.219E+03	-3.699E+06	-1.011E+04	-3.024E+03	-8.417E+02
	HH	1.764E+05	1.385E+04	5.843E+06	1.207E+04	5.007E+06
	Chi-square	7.261E-02	1.115E-02	1.241E-01	4.283E-02	7.025E-01
	R-square	0.8251				
Stage II	T-Phi	16.784	N/A	N/A	N/A	N/A
			13.964	16.888		
	TL	-10.535	N/A	N/A	N/A	N/A
	TH	53.795	N/A	N/A	N/A	N/A
	rho-Phi	0.2446	N/A	N/A	N/A	N/A
	HA	3.916E+04	N/A	N/A	N/A	N/A
	HL	-1.052E+05	N/A	N/A	N/A	N/A
	HH	9.623E+04	N/A	N/A	N/A	N/A
	Chi-square	2.653E-02	N/A	N/A	N/A	N/A
	R-square	0.9083				
Stage III	T-Phi	16.698	5.648	37.059	15.406	47.884
			16.200	16.997		
	TL	13.630	-157.033	165.213	5.111	1703.405
	TH	17.599	-108.049	229.838	-340748.550	22.908
	rho-Phi	0.1122	0.0897	4.7534	0.0918	7.7974
	HA	2.487E+04	1.946E+04	4.193E+04	1.707E+04	3.657E+04
	HL	-9.562E+05	-2.785E+06	-2.296E+03	-6.623E+06	-1.758E+05
	HH	3.543E+06	1.300E+04	5.729E+06	1.479E+06	1.913E+07
	Chi-square	7.052E-03	2.744E-04	1.815E-02	1.597E-03	2.141E-02
	R-square	0.6564				
Stage IV	T-Phi	-28.053	N/A	N/A	N/A	N/A
			-49.561	113.625		
	TL	54.007	N/A	N/A	N/A	N/A
	TH	-100.987	N/A	N/A	N/A	N/A
	rho-Phi	2.5219	N/A	N/A	N/A	N/A
	HA	9.969E+03	N/A	N/A	N/A	N/A
	HL	-9.309E+03	N/A	N/A	N/A	N/A
	HH	6.014E+03	N/A	N/A	N/A	N/A
	Chi-square	1.541E-03	N/A	N/A	N/A	N/A
	R-square	0.2141				

Notes: LCI and UCI = lower and upper 95% C.I.s estimated with the bootstrap percentile (BP), BC$_a$, or ABC methods; ABC-derived 95% C.I.s were calculated for T-Phi only, and are presented in the same columns as BP-derived 95% C.I.s, but one row beneath them; aE±b = a × 10$^{\pm b}$; N/A = could not be estimated; NaN = non-numeric estimate; T-Phi, TL, and TH estimates are presented in units of °C; for the units and meanings of all SSI model parameters, see equation (1) in the Introduction, and for their abbreviations in this Table and in *OptimSSI* output, see section 2.2.

For the majority of study and stage datasets to which the SSI model was fit, a reasonable fit with an $R^2 \geq 0.9$ was achieved (Tables 2-8), and a development curve of the appropriate shape corresponding to the SSI model over the range of biologically relevant temperatures for lobster larvae (ca. 0-40°C) was generated (Figures 1-6). Studies that considered a narrower range of test temperatures produced curves that were more narrow, whereas studies that tested a broader range of temperatures produced wider curves, as would be expected (Table 1; Figures 1-6).

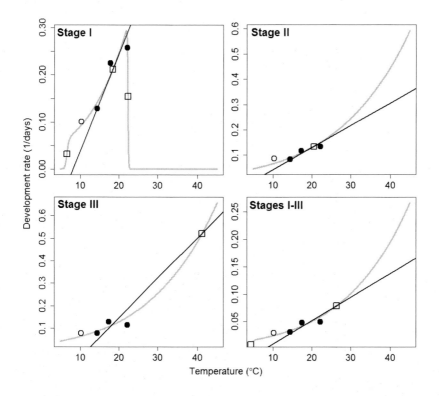

Figure 4. SSI model curves fit to development rates (y-axes, in 1/days) of American lobster larvae through different stages (different panels) at different temperatures (x-axes, in °C), using data obtained from Quinn et al. (2013). Solid circles = observed data within the linear portion of the development curve; open circles = observed data outside the linear range; gray lines = the SSI development curve fit to all observed data; black lines = linear fit of data in the linear portion of the development curve; open squares = estimates of (from left to right) T_L, T_Φ, and T_H based on the plotted data and curves. All plots made in R with the *SSIPlot* function of Ikemoto et al. (2013).

Table 5. SSI parameter estimates for data from Quinn et al. (2013)

	Parameter	Estimate	LCI (BP/ABC)	UCI (BP/ABC)	LCI (BC$_a$)	UCI (BC$_a$)
Stage I	T-Phi	18.438	15.155	20.408	15.806	20.408
			15.822	18.442		
	TL	6.621	-141.611	13.554	-141.611	13.554
	TH	22.349	22.349	366.667	NaN	NaN
	rho-Phi	0.2117	0.1500	0.2447	0.1634	0.2447
	HA	1.592E+04	5.419E+03	2.725E+04	5.419E+03	2.725E+04
	HL	-6.125E+05	-6.125E+05	-9.095E+03	NaN	NaN
	HH	2.007E+06	2.300E+04	2.007E+06	4.988E+05	2.007E+06
	Chi-square	5.111E-03	2.394E-12	3.742E-02	3.725E-10	3.742E-02
	R-square	0.9478				
Stage II	T-Phi	20.503	13.854	20.503	18.429	20.503
			20.503	21.411		
	TL	-132.717	-132.717	13.575	NaN	NaN
	TH	71.386	22.400	116.714	24.643	116.714
	rho-Phi	0.1341	0.0863	0.1382	0.0998	0.1382
	HA	1.067E+04	5.068E+03	1.943E+04	5.068E+03	1.943E+04
	HL	-1.239E+04	-5.828E+05	-1.239E+04	-7.244E+04	-1.239E+04
	HH	9.979E+04	5.279E+04	1.577E+06	4.798E+04	1.253E+05
	Chi-square	8.585E-03	1.898E-13	5.392E-02	1.898E-13	5.392E-02
	R-square	0.5845				
Stage III	T-Phi	41.172	11.269	51.897	15.647	51.897
			21.265	41.310		
	TL	-18.797	-141.659	13.710	-145.280	3.724
	TH	71.273	23.108	302.101	23.801	302.101
	rho-Phi	0.5213	0.0823	0.7523	0.0871	0.7523
	HA	1.151E+04	4.400E+03	2.855E+04	4.400E+03	2.855E+04
	HL	-9.766E+04	-6.582E+05	-1.221E+03	-6.582E+05	-4.094E+03
	HH	2.708E+05	3.603E+04	1.672E+06	4.834E+04	1.712E+06
	Chi-square	1.762E-02	1.244E-12	8.547E-02	7.777E-12	8.547E-02
	R-square	-0.0092				

Notes: LCI and UCI = lower and upper 95% C.I.s estimated with the bootstrap percentile (BP), BC$_a$, or ABC methods; ABC-derived 95% C.I.s were calculated for T-Phi only, and are presented in the same columns as BP-derived 95% C.I.s, but one row beneath them; aE±b = a × 10$^{\pm b}$; N/A = could not be estimated; NaN = non-numeric estimate; T-Phi, TL, and TH estimates are presented in units of °C; for the units and meanings of all SSI model parameters, see equation (1) in the Introduction, and for their abbreviations in this Table and in *OptimSSI* output, see section 2.2.

Table 6. SSI parameter estimates for data from Harrington et al. (2019)

	Parameter	Estimate	LCI (BP/ABC)	UCI (BP/ABC)	LCI (BC$_a$)	UCI (BC$_a$)
Stage I	T-Phi	20.987	14.150	21.486	14.678	21.486
			17.913	21.991		
	TL	3.487	-61.104	12.030	-61.104	7.352
	TH	53.553	18.334	86.202	24.267	86.202
	rho-Phi	0.2415	0.1281	0.2553	0.1322	2.55E-01
	HA	1.619E+04	1.187E+04	1.992E+04	1.187E+04	1.992E+04
	HL	-3.105E+05	-5.638E+05	-5.957E+03	-5.638E+05	-1.832E+05
	HH	1.943E+05	2.004E+04	1.763E+06	3.295E+04	1.763E+06
	Chi-square	1.000E-03	7.961E-12	1.472E-03	6.253E-11	1.472E-03
	R-square	0.9865				
Stage II	T-Phi	16.928	0.389	19.580	-24.925	18.815
			16.740	17.860		
	TL	13.401	-59.558	15.840	-22.294	15.840
	TH	22.504	22.065	435.758	21.487	29.521
	rho-Phi	0.1937	0.1028	0.3237	0.0976	2.64E-01
	HA	2.928E+04	2.122E+03	6.290E+04	2.122E+03	6.290E+04
	HL	-4.108E+05	-1.624E+06	-9.037E+03	-1.624E+06	-8.724E+04
	HH	2.534E+05	1.555E+04	2.483E+06	1.555E+04	2.483E+06
	Chi-square	2.298E-02	1.637E-12	2.874E-02	7.761E-09	2.874E-02
	R-square	0.8642				
Stage III	T-Phi	17.867	14.116	19.098	14.261	19.098
			17.210	19.015		
	TL	12.129	-127.299	15.506	-11.093	15.506
	TH	23.395	22.513	498.116	21.737	31.391
	rho-Phi	0.1548	0.0766	0.1831	0.0776	1.83E-01
	HA	2.447E+04	1.277E+04	4.039E+04	1.277E+04	3.458E+04
	HL	-1.718E+05	-6.985E+05	-9.224E+03	-6.985E+05	-5.333E+04
	HH	1.882E+05	1.214E+04	7.530E+05	2.600E+04	7.530E+05
	Chi-square	2.140E-03	5.282E-12	2.140E-03	1.043E-03	2.140E-03
	R-square	0.9789				

Notes: LCI and UCI = lower and upper 95% C.I.s estimated with the bootstrap percentile (BP), BC$_a$, or ABC methods; ABC-derived 95% C.I.s were calculated for T-Phi only, and are presented in the same columns as BP-derived 95% C.I.s, but one row beneath them; aE±b = a × 10$^{\pm b}$; N/A = could not be estimated; NaN = non-numeric estimate; T-Phi, TL, and TH estimates are presented in units of °C; for the units and meanings of all SSI model parameters, see equation (1) in the Introduction, and for their abbreviations in this Table and in *OptimSSI* output, see section 2.2.

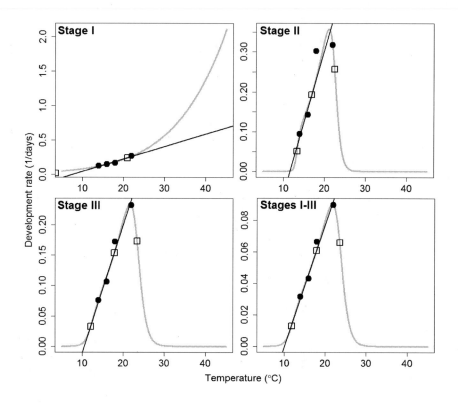

Figure 5. SSI model curves fit to development rates (y-axes, in 1/days) of American lobster larvae through different stages (different panels) at different temperatures (x-axes, in °C), using data obtained from Harrington et al. (2019). Solid circles = observed data within the linear portion of the development curve; open circles = observed data outside the linear range; gray lines = the SSI development curve fit to all observed data; black lines = linear fit of data in the linear portion of the development curve; open squares = estimates of (from left to right) T_L, T_Φ, and T_H based on the plotted data and curves. All plots made in R with the *SSIPlot* function of Ikemoto et al. (2013).

However, there were several cases in which the development curve generated was highly abnormal and unlikely to be realistic. In 8 cases, the data to which the model were fitted ended up being treated as located very early in the linear portion of the temperature-development curve, with the temperature at which the maximum development rate occurs and above which the rate declines until the T_H is reached being extrapolated to extremely high temperatures. This problem occurred for data for stages IV and I-IV from Templeman (1936) (Figure 1), stages II and I-III from Hudon and Fradette (1988) (Figure 3), stages II, III, and I-III from Quinn et al. (2013) (Figure 4), and stage I from Harrington et al. (2019) (Figure 5). Further, in addition to suffering the aforementioned problem data for stage IV from Hudon and Fradette (1988) also produced an inverted linear fit (Figure 3) to produce starting values to estimate T_L and T_H, resulting in extremely high T_L and extremely low T_Φ and T_H estimates for this dataset (Table 4). Therefore, the estimates of T_Φ, T_L, and T_H produced from such datasets are highly suspect and certainly incorrect. The structure of these datasets also in many cases did not permit bootstrap 95% C.I.s to be estimated, or if they were estimated they were often extremely wide and/or highly asymmetrical, limiting the ability for such estimates to be compared across studies or stages (Tables 4 and 5).

The average estimated value of T_Φ for lobster larvae across all study datasets was 17.44°C (range: 14.72-20.99°C) for stage I, 17.72°C (range: 16.78-20.50°C) for stage II, 21.81°C (range: 16.70-41.17°C) for stage III, 1.44°C (range: -28.05-18.00°C) for stage IV, 18.46°C (range: 13.70-26.33°C) for stages I-III combined, and 15.08°C (range: 14.17-16.373°C) for stages I-IV combined (Figure 8). The extremely low and unrealistic average estimated T_Φ for stage IV resulted from the highly abnormal development curve estimated for this stage from Hudon and Fradette's (1988) data (see above, and Figure 3). Three studies' datasets produced biologically unrealistic estimates of T_Φ: -28.053°C for stage IV (Hudon and Fradette (1988): Table 4 and Figure 3), 26.333°C for stages I-III combined (Quinn et al. (2013): Table 5 and Figure 4), and 41.172°C for stage III (Quinn et al. (2013): Table 5 and Figure 4). Aside from these extremes, estimates of T_Φ for American lobster larvae generally fell within the more biologically realistic range of 13.703-20.503°C (Tables 2-8).

Table 7. SSI parameter estimates for combined stages I-III data from various studies

	Parameter	Estimate	LCI (BP/ABC)	UCI (BP/ABC)	LCI (BC$_a$)	UCI (BC$_a$)
Hadley (1906)	T-Phi	18.368	16.014	18.776	16.014	18.776
			17.834	18.420		
	TL	14.040	4.030	14.040	9.383	14.040
	TH	24.216	24.084	33.323	24.084	24.232
	rho-Phi	0.0629	0.0456	0.0660	0.0456	0.0660
	HA	1.940E+04	1.643E+04	2.381E+04	1.643E+04	1.949E+04
	HL	-2.937E+05	-2.937E+05	-6.953E+04	NaN	NaN
	HH	2.160E+05	4.065E+04	2.160E+05	8.899E+04	2.160E+05
	Chi-square	5.543E-07	3.700E-12	5.543E-07	5.284E-12	5.543E-07
	R-square	1.0000				
Templeman (1936)	T-Phi	18.098	14.079	19.584	15.164	22.240
			18.076	18.139		
	TL	7.099	5.501	8.933	5.054	8.688
	TH	26.238	19.498	67.126	13.318	43.816
	rho-Phi	0.0616	0.0385	0.0707	0.0423	0.0833
	HA	1.821E+04	1.570E+04	2.316E+04	1.423E+04	2.265E+04
	HL	-6.595E+04	-1.652E+05	-5.496E+04	-9.950E+04	-1.217E+04
	HH	1.051E+05	1.373E+04	2.277E+05	1.724E+04	3.326E+05
	Chi-square	1.210E-03	7.244E-05	1.733E-03	7.092E-04	2.757E-03
	R-square	0.9916				
Hughes and Matthiessen (1962)	T-Phi	17.239	16.493	18.357	16.182	17.984
			17.131	17.418		
	TL	6.086	-33.519	14.561	-41.178	14.208
	TH	23.838	21.246	34.841	22.396	777.187
	rho-Phi	0.0486	0.0417	0.0634	0.0397	0.0588
	HA	3.390E+04	2.960E+04	4.201E+04	2.898E+04	4.025E+04
	HL	-7.760E+04	-7.841E+05	-1.270E+04	-3.983E+05	-6.781E+03
	HH	1.579E+05	7.663E+04	8.032E+05	5.296E+04	6.237E+05
	Chi-square	3.694E-02	1.713E-02	5.639E-02	2.331E-02	7.330E-02
	R-square	0.8543				
Ford et al. (1979)	T-Phi	20.012	3.610	23.727	5.482	28.689
			18.551	20.564		
	TL	12.988	-29.860	18.509	-2.142	4043.622
	TH	27.442	3.333	38.263	24.304	90.965
	rho-Phi	0.0678	0.0519	0.9470	0.0518	0.8684
	HA	1.302E+04	6.958E+03	8.297E+04	6.958E+03	1.544E+04
	HL	-1.619E+05	-3.306E+06	-2.826E+04	-3.306E+06	-2.826E+04
	HH	1.606E+05	4.987E+04	4.139E+06	4.067E+04	5.201E+05
	Chi-square	1.069E-03	1.050E-12	1.713E-03	1.020E-08	2.583E-03
	R-square	0.9108				
MacKenzie (1988)	T-Phi	15.972	12.881	18.952	12.871	18.768
			14.615	16.003		
	TL	8.078	4.494	10.910	-0.076	9.309
	TH	26.491	19.839	33.786	21.926	41.598

Study	Parameter	Estimate	LCI (BP/ABC)	UCI (BP/ABC)	LCI (BC$_a$)	UCI (BC$_a$)
	rho-Phi	0.0565	0.0339	0.0781	0.0338	0.0766
	HA	2.210E+04	1.723E+04	2.805E+04	1.723E+04	2.805E+04
	HL	-8.073E+04	-1.800E+05	-4.328E+04	-1.195E+05	-3.295E+04
	HH	5.958E+04	3.152E+04	1.412E+05	2.183E+04	1.069E+05
	Chi-square	1.253E-06	1.446E-11	1.253E-06	1.297E-07	2.727E-06
	R-square	1.0000				
Hudon and Fradette (1988)	T-Phi	13.703	13.123 / 13.663	16.897 / 17.296	12.919	16.657
	TL	6.571	-150.506	11.999	-1.025	14.637
	TH	67.413	17.731	337.263	25.868	1457.406
	rho-Phi	0.0325	0.0279	0.0616	0.0260	0.0567
	HA	3.317E+04	2.894E+04	3.891E+04	2.826E+04	3.757E+04
	HL	-1.449E+05	-5.333E+05	-1.022E+04	-4.664E+06	-7.887E+04
	HH	1.530E+04	1.302E+04	5.122E+05	3.716E+02	4.117E+04
	Chi-square	2.127E-03	4.516E-04	3.135E-03	1.171E-03	4.306E-03
	R-square	0.9408				
Quinn et al. (2013)	T-Phi	26.333	-91.320 / 21.957	35.598 / 31.809	16.478	35.598
	TL	4.437	-153.003	14.145	-153.003	14.145
	TH	114.756	22.364	755.530	23.493	755.530
	rho-Phi	0.0788	0.0085	0.0788	0.0483	
	HA	1.174E+04	1.268E+03	2.444E+04	1.268E+03	
	HL	-5.451E+05	-2.697E+06	-2.141E+03	-2.697E+06	
	HH	1.858E+05	1.041E+04	3.232E+06	1.031E+04	
	Chi-square	3.501E-03	3.149E-13	1.954E-02	1.756E-12	
	R-square	0.5796				
Harrington et al. (2019)	T-Phi	17.919	14.336 / 17.308	19.171 / 17.949	14.550	19.171
	TL	11.881	-146.901	15.565	-4.796	15.565
	TH	23.558	21.967	67.928	20.575	34.637
	rho-Phi	0.0612	0.0336	0.0730	0.0337	7.30E-02
	HA	2.284E+04	1.285E+04	3.648E+04	1.285E+04	3.648E+04
	HL	-1.632E+05	-7.560E+05	-1.015E+04	-7.560E+05	-6.119E+04
	HH	1.859E+05	2.957E+04	8.784E+05	3.240E+04	8.784E+05
	Chi-square	6.014E-04	1.691E-12	6.014E-04	1.888E-04	6.014E-04
	R-square	0.9831				

Notes: LCI and UCI = lower and upper 95% C.I.s estimated with the bootstrap percentile (BP), BC$_a$, or ABC methods; ABC-derived 95% C.I.s were calculated for T-Phi only, and are presented in the same columns as BP-derived 95% C.I.s, but one row beneath them; aE±b = a × $10^{\pm b}$; N/A = could not be estimated; NaN = non-numeric estimate; T-Phi, TL, and TH estimates are presented in units of °C; for the units and meanings of all SSI model parameters, see equation (1) in the Introduction, and for their abbreviations in this Table and in *OptimSSI* output, see section 2.2.

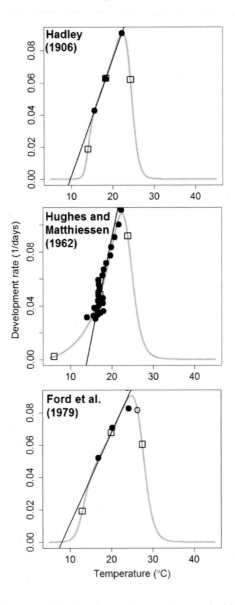

Figure 6. SSI model curves fit to development rates (y-axes, in 1/days) of American lobster larvae through stages I-III combined at different temperatures (x-axes, in °C), using data from three different studies (different panels). Solid circles = observed data within the linear portion of the development curve; open circles = observed data outside the linear range; gray lines = the SSI development curve fit to all observed data; black lines = linear fit of data in the linear portion of the development curve; open squares = estimates of (from left to right) T_L, T_Φ, and T_H based on the plotted data and curves. All plots made in R with the *SSIPlot* function of Ikemoto et al. (2013).

Table 8. SSI parameter estimates for combined stages I-IV data from various studies

	Parameter	Estimate	LCI (BP/ABC)	UCI (BP/ABC)	LCI (BC$_a$)	UCI (BC$_a$)
Templeman (1936)	T-Phi	16.373	13.068	18.736	13.492	18.946
			16.125	18.315		
	TL	6.302	-2.575	9.822	-192.688	9.141
	TH	73.779	15.589	238.965	32.371	134721.450
	rho-Phi	0.0259	0.0172	0.0335	0.0180	0.0345
	HA	1.963E+04	1.486E+04	2.433E+04	1.535E+04	2.451E+04
	HL	-8.902E+04	-7.594E+05	-4.606E+04	-3.283E+05	-9.551E+03
	HH	1.259E+04	5.203E+03	7.755E+05	2.691E+01	2.684E+04
	Chi-square	5.639E-04	1.685E-06	7.768E-04	3.282E-04	7.222E-03
	R-square	0.9831				
MacKenzie (1988)	T-Phi	14.700	12.939	19.946	12.510	16.266
			14.526	14.769		
	TL	9.066	3.430	11.199	5.664	11.199
	TH	34.083	17.951	53.108	20.307	78.910
	rho-Phi	0.0220	0.0175	0.0354	0.0164	0.0276
	HA	1.709E+04	1.394E+04	2.264E+04	1.394E+04	2.264E+04
	HL	-1.495E+05	-5.207E+05	-7.393E+04	-4.819E+05	-6.466E+04
	HH	3.383E+04	1.684E+04	4.959E+05	1.558E+04	1.316E+05
	Chi-square	4.240E-05	1.568E-12	5.277E-05	8.424E-06	6.054E-05
	R-square	0.9983				
Hudon and Fradette (1988)	T-Phi	14.169	13.627	16.027	12.904	14.797
			14.059	15.795		
	TL	11.775	-62.283	13.040	8.431	15.072
	TH	19.445	17.154	27.849	17.268	37.392
	rho-Phi	0.0215	0.0201	0.0269	0.0180	0.0232
	HA	1.877E+04	1.698E+04	2.020E+04	1.721E+04	2.046E+04
	HL	-6.019E+05	-2.097E+06	-2.297E+04	-3.390E+06	-1.379E+05
	HH	2.531E+05	6.778E+04	2.920E+06	2.709E+04	1.471E+06
	Chi-square	9.317E-05	3.125E-06	1.461E-04	4.186E-05	2.478E-04
	R-square	0.9841				

Notes: LCI and UCI = lower and upper 95% C.I.s estimated with the bootstrap percentile (BP), BC$_a$, or ABC methods; ABC-derived 95% C.I.s were calculated for T-Phi only, and are presented in the same columns as BP-derived 95% C.I.s, but one row beneath them; aE±b = a × 10$^{\pm b}$; N/A = could not be estimated; NaN = non-numeric estimate; T-Phi, TL, and TH estimates are presented in units of °C; for the units and meanings of all SSI model parameters, see equation (1) in the Introduction, and for their abbreviations in this Table and in *OptimSSI* output, see section 2.2.

The average estimated value of T_L for lobster larvae across all study datasets was -2.77°C (range: -37.11-7.76°C) for stage I, -22.76°C (range: -132.72-8.10°C) for stage II, 4.57°C (range: -18.80-13.63°C) for stage III, 20.67°C (range: -1.25-54.01°C) for stage IV, 8.90°C (range: 4.44-14.04°C) for stages I-III combined, and 9.05°C (range: 6.30-11.78°C) for stages I-IV combined (Figure 8). Clearly, most of these average estimates are biologically unrealistic, as they are subzero for stages I and II, above temperatures at which successful development has been reported for stages I-III and I-IV combined, and far too high and near the presumed optimum for stage IV. These results likely reflect the high variability in these estimates among study datasets, which was due to many datasets being unable to generate reasonable estimates of T_L, or indeed to be fit with the SSI model at all (see above). Biologically realistic estimates of T_L (i.e., between 0-6.7°C) were found for only 7 out of 29 (24.1%) cases (Tables 2-8).

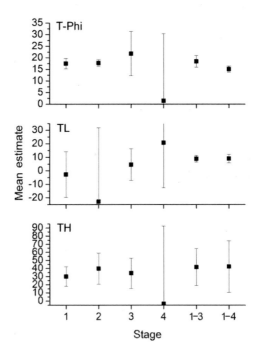

Figure 7. Estimates of the T_Φ (T-Phi), T_L (TL), and T_H (TH) parameters of the SSI model averaged across all study datasets (y-axes, in °C) for each American lobster larval stage or combination of stages (x-axes). Error bars represent ± 95% C.I.s calculated for each mean across study datasets (n = 3-8, depending on stage). Preliminary analyses were done in which each parameter (different panels) was compared across stages with separate one-way ANOVAs, but no significant differences were found (see section 3.1 for details).

The average estimated value of T_H for lobster larvae across all study datasets was 29.99°C (range: 20.39-53.55°C) for stage I, 39.79°C (range: 22.50-71.39°C) for stage II, 34.13°C (range: 17.60-71.27°C) for stage III, -3.48°C (range: -110.99-48.215°C) for stage IV, 41.74°C (range: 23.56-114.76°C) for stages I-III combined, and 42.44°C (range: 19.45-73.78°C) for stages I-IV combined (Figure 8). While most of these averaged estimates are more reasonable, they are quite high, exceeding ca. 40°C for stages II, I-III, and I-IV, and are thus unlikely to be real; further, the average estimate for stage IV is far too low (subzero). As for T_L estimates, high variability and unreliable parameter estimations from many study datasets was likely responsible for these unrealistic estimates. Biologically realistic estimates of T_H (i.e., those > 26.3°C and not much more than 40°C) were only generated in 8 out of 29 (27.6%) cases (Tables 2-8).

In preliminary analyses, when T_L, T_H, or T_Φ estimates were averaged across studies, there were some apparent differences among stages (Figure 7). Specifically, the T_H and T_Φ estimates were lower and the T_L estimates were higher for stage IV in comparison to those for all other larval stages; all estimates for stage IV also varied much more among studies than did those for other stages (Figure 7). Stage II also appeared to have a lower and more variable T_L than other stages (Figure 7). However, there were no statistically significant differences in these analyses among larval stages in their T_L (ANOVA: $F_{3,14} = 0.943$, p = 0.446), T_H (ANOVA: $F_{3,14} = 0.956$, p = 0.437), or T_Φ (ANOVA: $F_{3,14} = 2.126$, p = 0.143) estimates (Figure 7).

3.2. Comparisons among Larval Stages for Each Study

When data from Templeman (1936) were analyzed, the estimated intrinsic optimum temperature for larval development (T_Φ) of stage II was significantly lower than those estimated for all other stages, but the T_Φ of all other stages (I, III, and IV) did not significantly differ from one another (Table 2; Figure 8). The T_Φ was significantly higher for stage II than that for stage I when data from MacKenzie (1988) were assessed and 95% C.I.s were bootstrapped using the ABC method, while that for stage III did not significantly differ from those for these stages (Table 3). The estimated T_Φ for stage IV based on MacKenzie's data was lower than those for stages I-III, but could not be compared to them because its 95% C.I.s could not be bootstrapped using the ABC method (Table 3; Figure 8). The T_Φ values estimated with data from Hudon and Fradette (1988) did not significantly differ among larval stages (Table 4; Figure 8). When data from Quinn et al. (2013) were assessed, T_Φ values for all three larval stages significantly differed, with T_Φ values increasing with stage number (stage III > II > I) (Table 5; Figure 8). The T_Φ value estimated with data from Harrington et al. (2019) for stage I was significantly higher than that estimated for stage II; however, estimates for stages II and III did not significantly differ from one another, nor did those for stages I and III (Table 6; Figure 8).

When 95% C.I.s were bootstrapped using the bootstrap percentile or BC_a methods, the T_Φ values estimated for all datasets did not differ significantly among larval stages (Tables 2-6). Although the lower (T_L) and upper (T_H) threshold temperatures for larval development estimated for most study datasets both tended to increase with stage number, these did not significantly differ among larval stages based on comparisons using 95% C.I.s bootstrapped using the bootstrap percentile or BC_a methods (Tables 2-6).

3.3. Comparisons among Studies for Each Larval Stage

Comparisons of estimated values of the T_L, T_H, and T_Φ parameters among study datasets for the same larval stage or combinations of stages did not result in consistent overall differences among studies. Often estimates based on data from Hudon and Fradette (1988) (Tables 4, 7, and 8) and Quinn et al. (2013) (Tables 5 and 7) were lower or higher than those based on most other studies' data, but with no consistent directionality or significance (Tables 2, 3, and 6-8). Differences among studies were more often significant for the T_Φ than for the T_L and T_H parameters, especially when T_Φ estimates were compared based on 95% C.I.s bootstrapped using the ABC method (Tables 2-8; Figure 8).

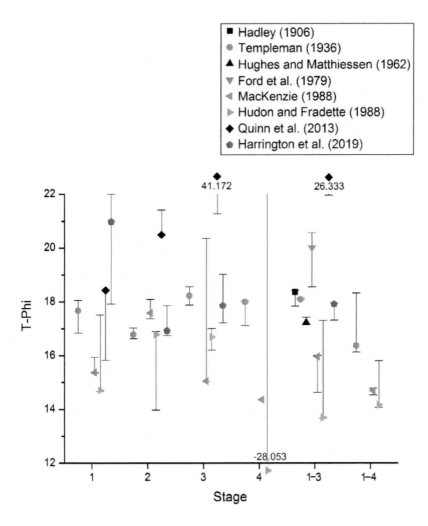

Figure 8. Estimates of the intrinsic optimum temperature (T_Φ or T-Phi) for the development of American lobster larvae estimated with the SSI model for each larval stage or combination of stages (x-axis) using data from different source studies (different symbol types; see Table 1). T-Phi estimates are presented in units of °C (y-axis). Error bars represent the lower and upper bootstrap 95% C.I.s calculated for each parameter estimate using the ABC method in R with the *mABCSSI* package of Ikemoto et al. (2013). For clarity, the y-axis in this figure is constrained to between 12-22°C; 95% C.I.s that extended beyond this range are not shown in their entirety, but the values of the three T-Phi estimates that were outside of this range are presented as numbers at the top or bottom of the graph. T-Phi estimates were compared among stages and studies based on whether their bootstrapped 95% C.I.s and estimates overlapped.

For analyses of T_Φ estimates for data for the total combined development from hatch to stage IV, the estimates based on different studies' data were all significantly different from one another based on their ABC-derived 95% C.I.s (Table 7; Figure 8), with the T_Φ estimated for Quinn et al. (2013) being significantly higher than that estimated for all other studies, and that for Harrington et al. (2019) being significantly higher than those for all other studies except for Quinn et al. (2013) (Table 7; Figure 8). For total development from hatch to stage V, the T_Φ values estimated for data from MacKenzie (1988) and Hudon and Fradette (1988) did not significantly differ from each other according to their ABC-derived 95% C.I.s, but that estimated for data from Templeman (1936) was significantly higher than those for both of these studies (Table 8; Figure 8).

3.4. Results for All Datasets Combined

The SSI model parameter estimates and plots of the SSI development function produced from analyses of all studies' datasets combined are presented in Table 9 and Figure 9, respectively.

If data from all studies were combined, the data for all stages produced reasonable development curves (Figure 9). The T_Φ value estimated for stage I based on data from all studies was significantly lower than those of stages II and III, but all other comparisons of T_Φ among stages were non-significant (Table 9). Comparisons of T_L and T_H estimates based on data from all studies combined among stages based on their bootstrap percentile- and BC_a-derived 95% C.I.s were all non-significant (Table 9), and T_Φ estimates also did not significantly differ among stages when assessed based on these bootstrapped 95% C.I.s (Table 9). Aside from the significant difference between the T_Φ of stage I and those of stages II and III, these comparisons of parameter estimates for all studies' data based on bootstrap 95% C.I.s generally agreed with the conclusions of preliminary analyses using ANOVAs (section 3.1 and Figure 7).

Table 9. SSI parameter estimates for data from all studies combined

	Parameter	Estimate	LCI (BP/ABC)	UCI (BP/ABC)	LCI (BC$_a$)	UCI (BC$_a$)
Stage I	T-Phi	16.534	15.087	18.279	13.836	17.334
			16.491	16.595		
	TL	4.936	-28.185	7.997	-22.081	8.384
	TH	23.883	21.581	25.506	22.057	745.066
	rho-Phi	0.212	0.172	0.267	0.136	0.239
	HA	2.254E+04	1.809E+04	2.982E+04	1.786E+04	2.944E+04
	HL	-6.018E+04	-2.715E+05	-2.070E+04	-1.029E+05	-9.947E+03
	HH	1.169E+05	9.241E+04	6.045E+05	1.063E+04	2.216E+05
	Chi-square	3.334E-01	1.309E-01	4.951E-01	2.198E-01	6.263E-01
	R-square	0.759				
Stage II	T-Phi	16.964	15.269	20.865	15.337	21.007
			16.874	17.001		
	TL	5.175	-105.491	8.731	-190.116	7.722
	TH	23.483	21.409	24.980	21.897	139.962
	rho-Phi	0.213	0.156	0.437	0.164	0.490
	HA	2.509E+04	2.053E+04	3.404E+04	2.038E+04	3.335E+04
	HL	-4.970E+04	-7.634E+05	-3.761E+03	-1.096E+05	-1.814E+01
	HH	1.186E+05	8.783E+04	1.535E+06	2.331E+03	3.260E+05
	Chi-square	3.728E-01	1.232E-01	6.411E-01	2.164E-01	1.215E+00
	R-square	0.688				
Stage III	T-Phi	16.903	15.362	22.498	14.238	18.975
			16.148	20.371		
	TL	5.314	-230.601	9.119	-138.502	9.885
	TH	25.145	22.197	62.618	22.233	1731.213
	rho-Phi	0.121	0.098	0.269	0.084	0.165
	HA	2.041E+04	1.607E+04	2.531E+04	1.498E+04	2.459E+04
	HL	-8.649E+04	-1.769E+06	-8.389E+02	-7.226E+06	-1.302E+04
	HH	1.404E+05	3.162E+04	3.570E+06	1.930E+01	1.074E+06
	Chi-square	1.139E-01	2.593E-02	1.981E-01	5.783E-02	2.753E-01
	R-square	0.821				
Stage IV	T-Phi	16.965	14.792	34.580	14.575	29.177
			16.355	17.618		
	TL	6.949	-3.920	9.560	-51.163	9.344
	TH	24.041	18.225	92.692	18.363	130.129
	rho-Phi	0.0562	0.0444	0.2279	0.0434	0.1760
	HA	1.479E+04	1.161E+04	1.824E+04	1.065E+04	1.788E+04
	HL	-7.394E+04	-2.982E+05	-3.286E+04	-1.809E+05	-5.753E+03
	HH	1.235E+05	1.990E+04	7.927E+05	1.220E+02	3.221E+05
	Chi-square	1.146E-02	2.608E-03	1.842E-02	6.161E-03	2.451E-02
	R-square	0.9043				

Table 9. (Continued)

	Parameter	Estimate	LCI (BP/ABC)	UCI (BP/ABC)	LCI (BC$_a$)	UCI (BC$_a$)
Stage I-III	T-Phi	17.326	16.157	18.879	16.326	19.878
			17.085	17.682		
	TL	0.095	-30.387	6.401	-48.507	5.962
	TH	25.661	24.264	26.173	24.590	28.581
	rho-Phi	0.055	0.046	0.068	0.048	0.080
	HA	2.254E+04	2.003E+04	2.620E+04	1.973E+04	2.576E+04
	HL	-3.718E+04	-1.599E+05	-1.393E+04	-9.960E+04	-7.230E+03
	HH	1.057E+05	9.246E+04	2.937E+05	8.296E+04	1.438E+05
	Chi-square	1.089E-01	6.012E-02	1.585E-01	7.240E-02	1.819E-01
	R-square	0.819				
Stage I-IV	T-Phi	16.192	14.331	18.038	14.800	18.742
			15.266	16.735		
	TL	7.804	6.619	9.242	5.928	9.077
	TH	25.526	19.121	31.856	20.035	40.534
	rho-Phi	0.0264	0.0211	0.0317	0.0223	0.0338
	HA	1.793E+04	1.614E+04	2.031E+04	1.578E+04	1.999E+04
	HL	-9.155E+04	-2.010E+05	-7.540E+04	-1.540E+05	-6.630E+04
	HH	8.639E+04	4.612E+04	3.732E+05	2.932E+04	2.541E+05
	Chi-square	1.716E-03	4.193E-04	2.604E-03	1.001E-03	3.787E-03
	R-square	0.9770				

Notes: LCI and UCI = lower and upper 95% C.I.s estimated with the bootstrap percentile (BP), BC$_a$, or ABC methods; ABC-derived 95% C.I.s were calculated for T-Phi only, and are presented in the same columns as BP-derived 95% C.I.s, but one row beneath them; aE±b = a × 10$^{±b}$; N/A = could not be estimated; NaN = non-numeric estimate; T-Phi, TL, and TH estimates are presented in units of °C; for the units and meanings of all SSI model parameters, see equation (1) in the Introduction, and for their abbreviations in this Table and in *OptimSSI* output, see section 2.2.

Based on all studies' data, T_Φ, T_L, and T_H estimates ranged from 16.534 to 17.326°C, 0.095 to 7.804°C, and 23.483 to 25.661°C, respectively (Table 9). All of these estimates of T_Φ, most estimates of T_L (except those for stages IV and I-IV, which were too high, 6.949 and 7.804°C, respectively), and none of the estimates of T_H (all were < 26.3°C) were biologically realistic (Table 9; Figure 9). Compared with estimates based on individual studies' datasets (Tables 2-8), those based on all studies combined were more consistent among stages, were more reliable in that they had smaller bootstrap 95% C.I.s, and tended to be more realistic (Table 9).

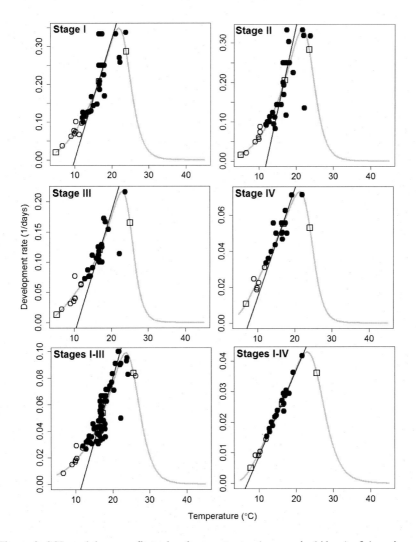

Figure 9. SSI model curves fit to development rates (y-axes, in 1/days) of American lobster larvae through different stages (different panels) at different temperatures (x-axes, in °C), using data from all studies combined (see Table 1). Solid circles = observed data within the linear portion of the development curve; open circles = observed data outside the linear range; gray lines = the SSI development curve fit to all observed data; black lines = linear fit of data in the linear portion of the development curve; open squares = estimates of (from left to right) T_L, T_Φ, and T_H based on the plotted data and curves. All plots made in R with the *SSIPlot* function of Ikemoto et al. (2013).

4. DISCUSSION

4.1. Variability and Issues among Estimates Based on Different Source Studies' Data

This chapter represents the first published application of the SSI development function to American lobster larval development data, as well as the first time that such a thermodynamic model has been used to estimate the intrinsic optimum temperature for the development of this species' larvae. When the estimated T_Φ, T_L, and T_H values for a given larval stage produced for data from different individual studies were compared, however, they were found to vary markedly from one another, and in many cases the biological realism of these estimates and their reliability (inferred based on the width of their bootstrap 95% C.I.s) was very poor. Therefore, the discussion in the present section was restricted to explaining the sources of these variations and issues with individual studies' results, while comparisons of thermal parameter estimates to the published literature and discussion of their implications were saved for the discussion of estimates based on all studies' data combined in the following section (4.2).

The source data came from studies that used different larval source populations, and thus some differences could be expected among studies as a result of geographic differences in temperature-dependent development reflecting local thermal adaptation or acclimation (Quinn et al. 2013; 2017); indeed, the clear differences between the shapes of the development curves (and consequently thermal parameters) derived for data from Quinn et al. (2013) from those for all other studies, which were conducted in more southerly and relatively warm-water regions (Quinn and Rochette 2015), may support the conclusion that that study found evidence of thermal adaptation. However, the magnitudes of the differences found herein in thermal parameters, including among geographically close source locations, were far too great for this be a sufficient explanation by itself.

The data analyzed came from eight different studies spanning more than a century of research on lobster larvae (1906-2019), which obviously varied in a number of additional ways from one another in terms of the temperature range tested, setting (lab vs. field), holding conditions of larvae (individual vs. communal), temperature control (constant vs. variable) (Table 1), and many other ways (e.g., food, photoperiod, sample sizes, etc.; reviewed by MacKenzie (1988) and Quinn et al. (2013)). Such differences among studies were thus more likely reasons for the observed variability, and thus the limiting and optimum temperatures estimated based on any one specific study herein should likely not be considered representative of the entire species' developmental characteristics. The field sampling study of Hudon and Fradette (1988) produced particularly problematic estimates, which was likely due to the limited range of temperatures that can be captured by field sampling (see below), as well as possible errors resulting from not tracking the same larval individuals or cohorts across sampling dates (Annis et al. 2007; Quinn et al. 2013, 2017; Gendron et al. in press). In the future, then, estimation of SSI model parameters based on such indirect field sampling methods should probably be avoided. The small samples sizes included in the data from most of the individual studies examined (Table 1) were also likely too small to permit optimal fitting with the SSI model and the calculation of very precise 95% C.I.s for its parameter estimates (Ikemoto et al. 2013); this justified the analyses performed on combined data (see section 4.2).

The factor that likely had the greatest impact on the differences in the estimated thermal parameters among studies was the difference in the thermal ranges they used (Kontopoulos et al. 2018; Table 1). Although overall a thermal range from 6.7 to 26.3°C was investigated, which should extend into the nonlinear portions of the temperature-development curve and approach the actual T_L and T_H values for lobster larvae, no single study covered this entire range, and some covered very small ranges (e.g., 3.2-6.2°C: Hudon and Fradette 1988) and/or only extended towards very low (Templeman 1936) or high (Ford et al. 1979) temperatures, but not both. The fitting of a complex nonlinear function like the SSI model to development data can only be accurate if the data cover both the lower and upper nonlinear portions of the development curve in addition to the intermediate linear portion (Schoolfield et al. 1981; Shi et al. 2011, 2017, 2019; Ikemoto et al. 2013). If one or more portions of this range are missing, then estimated thresholds will be placed either just outside of the observed data range (if there is some nonlinearity in the data) or very far beyond it (if the data are highly linear), which in either case results in unrealistic estimates (Jafari et al. 2012; Shi et al. 2012a, b, 2013; Quinn 2017b; Kontopoulos et al. 2018), and will also create quite wide and asymmetrical bootstrap 95% C.I.s. This can also result in a fit to data from one study estimating a threshold within the range of temperatures over which development was observed to be completed in another study; for example, the very high T_L estimate of 12.988°C for data from Ford et al. (1979), who only reared larvae at temperatures higher than 16.9°C, was well within the range of temperatures at which development was completed in most other studies considered (Table 1).

4.2. Interpretation and Implications of Estimates Based on All Studies' Data Combined

To account for the above shortcomings, in this chapter SSI fits were also attempted using all studies' data combined. Although doing this masked much of the variability within the datasets (e.g., the strong differences in the development curves of Quinn et al. (2013) from all others; Figure 9) and was still not able to estimate biologically realistic T_H values > 26.3°C (Table 9), it did allow the widest thermal range possible to be investigated. Further, this approach resulted in much more reasonable and consistent parameter estimates than did analyses with individual study datasets. It also allowed a few patterns to be detected that were missed by preliminary analyses using conventional ANOVAs to make comparisons among stages. Using this approach, the T_Φ, T_L, and T_H of American lobster larvae were estimated to be 16.534-17.326°C, 0.095-7.804°C, and 23.483-25.661°C, respectively (Table 9). Although these estimates did increase, to some extent, with stage number (i.e., thermal parameters of later stages tended to have higher values than those of earlier stages), none of these parameters differed significantly among stages except that the T_Φ of stage I was less than that of stages II and III.

The range of T_L values found was mostly reasonable, although a few were too high, while all T_H estimates were likely too low, but perhaps close to (i.e., within ~2°C of) being reasonable. The range of estimates of T_L and T_H produced agreed in general with previous experimental observations (Huntsman 1924; Sastry and Vargo 1977; Gruffydd et al. 1979; Quinn 2017a) and with those derived using other nonlinear development functions by Quinn (2017b) but with fewer T_L values of 0°C found, perhaps related to the inability for SSI parameter estimates to be constrained in the present analyses (see section 4.3 below). The lack of significant differences in T_L and T_H among stages may disagree with previous studies, in which thermal tolerance was suspected of increasing with stage number (e.g., Huntsman 1924; Gruffydd et al. 1979) or being lower in stage II (Sastry and Vargo 1977), although such results were overall inconclusive (Quinn 2017a). There is evidence from many arthropod taxa that developmental proportions, temperature effects, and thermal thresholds differ among stages (Quinn 2019). However, the rate isomorphy hypothesis developed in other studies of ectotherms, including arthropods like crustaceans (e.g., Jarošík et al. 2004), alternatively suggests that such thermal developmental threshold should not differ among stages within the same species, and that most apparent differences in these found among stages result from experimental errors. The lack of differences found herein therefore may agree with rate isomorphy, although due to the wide variability in the estimates obtained thus far this is uncertain.

Ultimately, if the thermal optima and limits for development differ among larval stages, this will have important implications to how larvae are affected by temperature changes in nature; for example, if the optima and tolerance limits of earlier stages are lower than those of later ones, then they could be impacted the soonest and most severely by climate warming (Sastry and Vargo 1977; Schmalenbach and Franke 2010; Caputi et al. 2013; Quinn 2019). Although the overall results found herein (as well as some of the results from analyses of individual study datasets) may suggest such stage-specific differences, the magnitudes and significances of the differences among the T_L and T_H estimates were too low and/or variable to conclude that such differences exist based on these data. However, The T_Φ

estimates produced through these analyses were much more consistent and realistic, and thus considering them may be more informative. According to Ikemoto et al. (2013), the T_Φ parameter in the SSI model is related to the other thermal parameters (e.g., T_L and T_H) therein as follows:

$$T_\Phi = \frac{\Delta H_L - \Delta H_H}{R \ln\left(-\frac{\Delta H_L}{\Delta H_H}\right) + \left(\frac{\Delta H_L}{T_L}\right) - \left(\frac{\Delta H_H}{T_H}\right)} \tag{2}$$

Therefore, comparisons of more reliable T_Φ estimates may allow inferences to be drawn about the general differences in thermal tolerances, if there are any, among lobster larval stages, including differences in their presumed T_L and T_H parameters.

The overall values and consistency among stages of the T_Φ estimates found herein were interesting. In previous research, the T_Φ of terrestrial insects and arachnids was found to be mainly around 20°C (ca. 15-25°C) (Ikemoto 2005, 2008; Shi et al. 2011, 2012a, 2012b, 2013; Jafari et al. 2012; Ikemoto and Egami 2013; Padmavathi et al. 2013; Sreedevi et al. 2013), while those for decapod crustaceans varied among habitats, being approximately 7-9°C or 19-27°C for species from cold- and warm-water habitats, respectively (Yamamoto et al. 2017) (Shi et al. 2019). The estimates found herein for American lobster are thus reasonable in light of the literature on arthropod developmental optima. Previous studies of American lobsters estimated (based on various criteria) that the optimal temperature for rearing their larvae was ca. 18-20°C (Van Olst et al. 1976; Carlberg and Van Olst 1977; Ford et al. 1979; Bartley et al. 1980; Quinn 2017a), which is close to (but slightly higher than) the intrinsic optimum temperatures estimated herein; these two sets of findings therefore lend some support to one another. The fact that, with one exception, the T_Φ was mostly consistent across stages also agreed with the rate isomorphy hypothesis (Jarošík et al. 2004), but not with other studies that suggested differences in thermal optima among developmental stages (e.g., Sastry and Vargo 1977; Quinn 2019). However, the significantly lower T_Φ of stage I than that of stages II and III does provide some evidence that thermal tolerances increase, if only slightly, among larval stages, which does agree somewhat with past studies (Huntsman 1924; Gruffydd et al. 1979) and make sense in light of the seasonal thermal trajectory from the time of larval release to settlement across most of the species' range (Ennis 1995; Lawton and Lavalli 1995; Quinn and Rochette 2015).

The implications of lobster larvae developing optimally between 16.534 and 17.326°C are also of interest. Throughout most of this species' range, seawater temperatures ≥ 18°C were historically uncommon during the summer, when larvae are in the water (Quinn and Rochette 2015; Jaini et al. 2018), but recent climate change has resulted in waters being warmer during the summer (Guo et al. 2013; Quinn 2017a). In historically colder parts of the species' range, warming has been associated with increased abundances and fisheries landings of lobsters (Caputi et al. 2013; Jaini et al. 2018), which may be partly associated with lobster larvae more often experiencing temperatures close to or at their intrinsic developmental optimum, in addition to there being beneficial impacts of warming on other life stages (adults, juveniles, and eggs) and processes (e.g., predator-prey interactions: Boudreau et al. 2015). Conversely, in historically warmer parts of the species' range, temperatures exceeding 22°C now occur in the summer, and lobster abundances have declined (Guo et al. 2013; Boudreau et al. 2015); perhaps the fact that lobster larvae are exposed to supra-optimal temperatures in these regions, resulting in stress, physiological impairment, and perhaps some mortality, is related to these recent declines (Quinn 2017a). Throughout the species' range, summer water temperatures experienced by larvae are expected to increase into the future (Guo et al. 2013; Quinn 2017a), so it is possible that more lobster larvae will experience temperatures too far above their optima, perhaps even approaching their T_H, which may lead to further declines. The possibly lower optimum temperature found for stage I than those for later stages could also mean that this stage will be affected first and most strongly by rising water temperatures in the future.

4.3. Findings Relevant to SSI Model Fitting Procedures

In this chapter, comparisons of T_Φ estimates based on their 95% C.I.s bootstrapped using the ABC method detected statistically significant differences among stages and studies far more often than equivalent comparisons based on 95% C.I.s bootstrapped using the bootstrap percentile and BC_a methods. This confirms the conclusion of Ikemoto et al. (2013) that it is most appropriate to bootstrap and compare C.I.s of T_Φ estimates using the ABC method. Congruently, all comparisons of T_L and T_H estimates, which were based on analyses of bootstrap percentile- or BC_a-derived C.I.s only, were non-significant. It seems reasonable to expect, however, that if 95% C.I.s could be calculated for T_L and T_H estimates using the ABC method, then some significant differences might be detected that are lost in comparisons using bootstrap percentile- or BC_a-derived C.I.s because these tend to innately be too wide and variable, particularly for limited datasets such as those used in the present chapter. Although the T_L and T_H of a species are related to its T_Φ (Ikemoto et al. 2013; see equation (2) above), it is still important to know their precise values when making predictions of future changes. Therefore, in the future, a follow-up study to Ikemoto et al. (2013) should be done to develop a program like their *mABCSSI* package that would allow bootstrap C.I.s for T_L and T_H estimates to be calculated using the ABC method.

Many of the thermal parameter estimates derived from the SSI model in the present chapter were also concluded to be biologically unrealistic. This resulted partially because the *OptimSSI* package does not currently contain the means to constrain its parameter estimates. When performing nonlinear regressions in many statistical packages, for example IBM SPSS Statistics, it is possible to define constraints on development function parameters that can 'force' them to be estimated within bounds defined by the user as 'realistic' (Guerrero et al. 1994); for example, this approach was used by Quinn (2017b) when fitting various complex nonlinear development functions including T_L and/or T_H parameters to arthropod larval development data. Typically this approach achieves a poorer fit than an unconstrained regression would, but avoids estimating unrealistic values for parameters that are defined as having biological meaning (Kontopoulos et al. 2018). Therefore, the addition of a means to constrain thermal parameter estimate values to the *OptimSSI* package (or more accurately a successor to it) should also be pursued.

The evaluation of the biological realism of T_L and T_H may also need to be revisited, or the implications of these parameters as defined in the SSI model to developmental observations considered more carefully. In the SSI model, T_L and T_H are the temperatures at which an enzyme involved in development is half active and half temperature-inactivated (Sharpe and DeMichele 1977; Schoolfield et al. 1981; Shi et al. 2011; Ikemoto et al. 2013), and thus at these temperatures not *all* copies of the enzyme are inactivated (Kontopoulos et al. 2018). Thus, while development would certainly be greatly impaired at or below these temperatures, it would not be entirely impossible. In laboratory studies, larval survival at near-extreme temperatures is often markedly reduced, but some individuals still manage to complete development (e.g., Templeman 1936; MacKenzie 1988). In this case, it is conceivable that they could be experiencing T_L or T_H conditions corresponding to the definitions in the SSI model, but development is still possible (Kontopoulos et al. 2018). This differs from the definition of equivalent lower and upper thermal thresholds in other development functions, at which development is supposed to be impossible (Quinn 2017b). Some of the estimated thermal limits produced in this chapter that were considered biologically unrealistic were only slightly within the range over which previous studies had observed some successful development, although with impaired larval survival (e.g., within 1-2°C of 6.7°C or 26.3°C (Templeman 1936; Ford et al. 1979)). Such estimates may thus in fact be realistic, but would need to be interpreted differently; this possibility should be investigated in a later study.

The definition of which portion of the datasets subjected to analyses with the *OptimSSI* package was within the 'linear' portion of the temperature-development curve also impacted the shapes of the development curves and values of the thermal parameter estimates it produced (*sensu* Campbell et al. 1974; Ikemoto and Takai 2000; Ikemoto et al. 2013). This was because the linear fit produced at this point provided starting values for the values of T_L (and thus indirectly of T_H and T_Φ, since these are interrelated (Ikemoto et al. 2013)) that were then fed into nonlinear regression procedures within this program, and such starting values can impact the final parameter estimates produced through such procedures (Campbell et al. 1974; Quinn 2017b). The data for stage IV analyzed from Hudon and Fradette (1988) herein provide an extreme example of this: due to the very narrow thermal range examined and the wide variability in development rates within it, the linear fit produced ended up having a *negative* slope (Figure 9), implying that development generally *slowed* as temperature increased, and resulting in the T_L produced being *greater* than the T_Φ and T_H (which were < 0°C!) (Table 4, Figure 8). In less extreme cases, the relatively shallow slope of the linear fit produced (e.g., data for Quinn et al. (2013)) and/or the lack or limited amount of data within the nonlinear portion of the development curve led to unnaturally extreme thermal parameter estimates and abnormal development curves (see sections 3.2 and 3.3).

In a series of side analyses (results not shown), for several of the datasets for which biologically unrealistic parameter estimates and/or abnormal development curves were produced using the definition of the linear portion set herein (temperatures > 12°C and < 26.3°C; see section 2.2), changing the bounds of the linear portion produced more reasonable results. However, this procedure of 'optimizing' the linear portion's definition for each dataset was avoided in the main results presented in this chapter because it was considered potentially dubious. As all datasets examined herein were for the same species, it was unclear whether it was justifiable to redefine the shape of the development curve from study to study and/or stage to stage to this extent. It is entirely possible that the shape of the temperature-development curve, including the breath and bounds of its linear portion, could differ among the different populations of the same species examined in different source studies due to geographic variation (e.g., Quinn et al. 2013). However, given that no study has fully spanned the entire range of temperatures over which American lobster larval development is possible (see above and Table 1), we do not currently have sufficient information to conclude exactly where transitions in the development curve occur, much less whether their thermal location differs geographically. The potential sensitivity of the *OptimSSI* package to the definition of the linear portion of the development curve and its implications will bear further investigation, as will the better definition of the full development curve(s) of lobster larvae.

4.4. Need for New Rearing Studies to be Analyzed with the SSI Model

The present chapter produced a large database of SSI model parameters and their bootstrap C.I.s that may be useful for future studies of this species. However, based on the wide variability of the results among stages and source studies, this information cannot, unfortunately, be said to provide conclusive evidence of the exact optimum and limiting temperatures for the development of lobster larvae (although results produced based on all studies' data combined (Table 9) are potentially promising). The chief issue confounding the results is the fact that no one study of American lobster larval development has yet spanned the entire range of temperatures biologically relevant to the larvae of this species, i.e., those extending from its actual T_L to its actual T_H, which must be < 6.7°C (Templeman 1936) and > 26.3°C (Ford et al. 1979), respectively (Quinn 2017a). However, this shortcoming is by no means unique to the American lobster. For the vast majority of marine decapod crustaceans, studies of larval development have historically focused on rearing larvae at a few temperatures within a limited, but presumably most biologically relevant, range to assess their differential survival, growth, development times, etc. at different temperatures (Anger 2001; Quinn 2017b). Studies of crustacean development at thermal extremes and the identification of thermal thresholds have thus been less of interest, and rearing over a wide range of temperatures including extremes is often impractical and expensive for marine larvae, while conversely such studies are commonplace for terrestrial insects and arachnids (Quinn 2017b and references therein; see also: Campbell et al. 1974; Ikemoto 2005, 2008; Shi et al. 2011, 2012a, 2012b, 2013; Jafari et al. 2012; Ikemoto and Egami 2013; Padmavathi et al. 2013; Sreedevi et al. 2013).

However, this is beginning to change (e.g., Yamamoto et al. 2017), and given concerns about the effects of climate change on American lobster fisheries recruitment (Caputi et al. 2013; Boudreau et al. 2015; Jaini et al. 2018; Quinn 2017a) it is time for such an extensive and thorough study to be done for this species' larvae. For example, a study rearing larvae in the laboratory at as many controlled temperatures as possible (optimally ≥ 10; Ikemoto et al. 2013) between ca. 0-5°C and ca. 27-30°C or higher should help to identify the lower and upper developmental thresholds and better define the shape of the development curve between them.

It should also be noted that most thermal parameter estimates produced for stage IV from a single study's dataset or with all studies' data combined in this chapter tended to be particularly variable and/or unrealistic. This was likely due in part to the low sample sizes available within individual studies for this stage, as well as the low number (3) of studies that provided data for its duration (Table 1). However, because stage IV is the settling stage of the American lobster's lifecycle, its duration can also be affected strongly by other factors besides temperature, especially the availability and type of bottom substrate. For example, in the absence of suitable substrate for settlement (cobble), this stage can be prolonged as a form of settlement delay, whereas if substrate is provided it can be greatly shortened (Phillips and Sastry 1980; Ennis 1995; Lawton and Lavalli 1995). Previous development studies done in the laboratory did not provide stage IV lobsters with substrate (Templeman 1936; MacKenzie 1988), which may have artificially prolonged this stage and also exacerbated developmental and behavioral differences among individuals. Therefore, a future study aiming to assess thermal parameters of stage IV development must also account for the impacts of substrate on its duration, for example by holding stage IV lobsters not only at different temperatures, but also with or without different types of substrate for settlement.

Once such a study has been done, analyzing its data using the SSI model, as done herein, should provide valuable information on the temperature-dependent development of this species' larvae, including their developmental thresholds and optimum temperature(s), which can then be used to predict the effects of future climate change on them. This chapter provides a framework to support the enactment of such a study. Further, the approaches taken herein could and likely should be applied to data for other species of lobsters and other decapod crustaceans, as these also support important fisheries that depend on larval supplies (e.g., Yao and Zhang, 2018) that are potentially impacted by changes in water temperature (Phillips and Sastry 1980; Caputi et al. 2013; see also other chapters in this volume).

CONCLUSION

The estimates of the thermal parameters (T_Φ, T_L, and T_H) of American lobster larval development obtained using the SSI model in this chapter varied widely among studies, but with no consistent and few significant differences in them among stages. Although most T_Φ estimates were biologically realistic, the majority of T_L and T_H estimates were not. When all studies' datasets were combined, the T_Φ, T_L, and T_H for American lobster larval development were estimated to be 16.534-17.326°C, 0.095-7.804°C, and 23.483-25.661°C, respectively. The T_Φ of stage I was significantly lower than those of stages II and III, but all other differences among stages were non-significant. However, none of the previous studies whose data were analyzed herein were conducted over a wide enough thermal range to allow for a conclusive fit of their data with the SSI model to be achieved. Therefore, a future rearing study conducted over a wide thermal range, followed by analyses with the SSI model, is needed to better define the limiting and optimum temperatures for the development of this species' larvae for use in predicting the potential effects of climate change on its fisheries recruitment.

ACKNOWLEDGMENTS

The University of New Brunswick (UNB), Saint John Campus provided access to resources needed for this research. I thank Peijian Shi for introducing me to the SSI model and the *OptimSSI* package, and also providing helpful advice and guidance in its use. Erin Miller reviewed this chapter and provided helpful comments that improved it. I also thank M. Quinn for regular help and support, and Nadya S. Columbus and Nova Science Publishers, Inc. for inviting me to contribute to the present collection.

REFERENCES

Amarasekare P, Savage V (2012) A framework for elucidating the temperature dependence of fitness. *Am Nat* 179:178-191.

Anger K (2001) *Crustacean Issues 12: The Biology of Decapod Crustacean Larvae*. Rotterdam, The Netherlands: A.A. Balkema.

Angilletta MJ Jr. (2006) Estimating and comparing thermal performance curves. *J Therm Biol* 31:541-545.

Annis ER, Incze LS, Wolff N, Steneck RS (2007) Estimates of *in situ* larval development time for lobster, *Homarus americanus*. *J Crustac Biol* 27:454-462.

Annis ER, Wilson CJ, Russell R, Yund PO (2013) Evidence for thermally mediated settlement in lobster larvae (*Homarus americanus*). *Can J Fish Aquat Sci* 70:1641–1649.

Bartley DM, Carlberg JM, Van Olst JC, Ford RF (1980) Growth and conversion efficiency of juvenile American lobsters (*Homarus americanus*) in relation to temperature and feeding level. *Proc World Maricul Soc* 11:355-368.

Bělehrádek J (1935) *Temperature and Living Matter: Protoplasma Monographia, No. 8*. Berlin, Germany: Borntraeger, 277 p.

Boudreau SA, Anderson SC, Worm B (2015) Top-down and bottom-up forces interact at thermal range extremes on American lobster. *J Anim Ecol* 84:840-850.

Brière JF, Pracross P, Rioux AY, Pierre JS (1999) A novel rate model of temperature-dependent development for arthropods. *Environ Entomol* 28:22-29.

Byrne M (2011) Impact of ocean warming and ocean acidification on marine invertebrate life history stages: vulnerabilities and potential for persistence in a changing ocean. *Oceanogr Mar Biol* 49:1-42.

Campbell A, Frazer BD, Gilbert N, Gutierrez AP, Mackauer M (1974) Temperature requirements of some aphids and their parasites. *J Appl Ecol* 11:431-438.

Caputi N, de Lestang S, Frusher S, Wahle RA (2013) The impact of climate change on exploited lobster stocks. In: Phillips BF (ed.) *Lobsters: Biology, Management, Aquaculture and Fisheries* (2nd Ed.). Oxford, UK: John Wiley and Sons Ltd., p. 84-112.

Carlberg JM, Van Olst JC (1977) Methods for culturing the American lobster (*Homarus americanus*). In: *3rd Meeting of the I.C.E.S. Working Group on Mariculture, Breat, France, May 10–13, 1977* (Actes de Colloques du C.N.E.X.O. 4), p. 261-275.

Corkrey R, Olley J, Ratkowsky D, McMeekin T, Ross T (2012) Universality of thermodynamic constants governing biological growth rates. *PLoS ONE* 7:e32003. DOI:10.1371/journal.pone. 0032003.

Cummings G, Fiddler F, Vaux DL (2007) Error bars in experimental biology. *J Cell Biol* 177:7-11.

DiCiccio TJ, Efron B (1992) More accurate confidence intervals in exponential families. *Biometrika* 79:231-245.

DiCiccio TJ, Efron B (1996) Bootstrap confidence intervals (with Discussion). *Stat Sci* 11:189-228.

Efron B, Tibshirani RJ (1993) *An Introduction to the Bootstrap*. New York, NY, USA: Chapman and Hall/CRC.

Ennis GP (1995) Larval and postlarval ecology. In: Factor JR (ed.) *Biology of the Lobster* Homarus americanus. New York, NY, USA: Academic Press Inc., p. 23-46.

Ford RF, Felix JR, Johnson RL, Carlberg JM, Van Olster JC (1979) Effects of fluctuating and constant temperatures and chemicals in thermal effluent on growth and survival of the American lobster (*Homarus americanus*). *Proc World Maricul Soc* 10:139-158.

Forster J, Hirst AG, Woodward G (2011) Growth and development rates have different thermal responses. *Am Nat* 178:668-678.

Gendron L, Lefaivre D, Sainte-Marie B (in press) Local egg production and larval losses to advection contribute to interannual and long-term variability of American lobster (*Homarus americanus*) settlement intensity. *Can J Fish Aquat Sci*. DOI:10.1139/cjfas-2017-0565.

Gruffydd LLD, Rieser RA, Machin D (1975) A comparison of growth and temperature tolerance in the larvae of the lobsters *Homarus gammarus* (L.) and *Homarus americanus* H. Milne Edwards (Decapoda, Nephropidae). *Crustaceana* 28:23-32.

Guerrero F, Blanco JM, Rodríguez V (1994) Temperature-dependent development in marine copepods: a comparative analysis of models. *J Plankton Res* 16:95-103.

Guo L, Perrie W, Long Z, Chassé J, Zhang Y, Huang A (2013) Dynamical downscaling over the Gulf of St. Lawrence using the Canadian Regional Climate Model. *Atmos Ocean* 51:265-283.

Hadley PB (1906) Regarding the rate of growth of the American lobster (*Homarus americanus*). *Annu Rep Comm Inland Fish RI* 36:156-235.

Harrington AM, Tudor MS, Reese HR, Bouchard DA, Hamlin HJ (2019) Effects of temperature on larval American lobster (*Homarus americanus*): is there a trade-off between growth rate and developmental stability? *Ecol Indicat* 96:404-411.

Hudon C, Fradette P (1988) Planktonic growth of larval lobster (*Homarus americanus*) off Iles de la Madeleine (Quebec), Gulf of St. Lawrence. *Can J Fish Aquat Sci* 45:868-878.

Hughes JT, Mathiessen GC (1962) Observations on the biology of the American lobster, *Homarus americanus*. *Limnol Oceanogr* 7:414-421.

Huntsman AG (1924) Limiting factors for marine animals 2. Resistance of larval lobsters to extremes of temperature. *Contrib Can Biol Fish* 2:89-94.

Ikemoto T (2005) Intrinsic optimum temperature for development of insects and mites. *Environ Entomol* 34:1377-1387.

Ikemoto T (2008) Tropical malaria does not mean hot environments. *J Med Entomol* 45:963–969.

Ikemoto T, Egami C (2013) Mathematical elucidation of the Kaufmann effect based on the thermodynamic SSI model. *Appl Entomol Zool* 48:313-323.

Ikemoto T, Kurahashi I, Shi PJ (2013) Confidence interval of intrinsic optimum temperature estimated using thermodynamic SSI model. *Insect Sci* 40:240-248.

Ikemoto T, Takai K (2000) A new linearized formula for the law of total effective temperature and the evaluation of line-fitting methods with both variables subject to error. *Environ Entomol* 29:671-682.

Jafari S, Fathipour Y, Faraji F (2012) Temperature-dependent development of *Neoseiulus barkeri* (Acari: Phytoseiidae) on *Tetranychus urticae* (Acari: Tetranychidae) at seven constant temperatures. *Insect Sci* 19:220-228.

Jaini M, Wahle RA, Thomas AC, Weatherbee R (2018) Spatial surface temperature correlates of American lobster (*Homarus americanus*) settlement in the Gulf of Maine and southern New England shelf. *Bull Mar Sci* 94:737-751.

Jarošík V, Kratochvíl L, Honek A, Dixon AFG (2004) A general rule for the dependence of developmental rate on temperature in ectothermic animals. *Proc Biol Sci* 271:S219–S221.

Jost JA, Podolski SM, Frederich M (2012) Enhancing thermal tolerance by eliminating the *pejus* range: a comparative study with three decapod species. *Mar Ecol Prog Ser* 444:263-274.

Kontopoulous D-G, García-Carrera B, Sal S, Smith TP, Pawar S (2018) Use and misuse of temperature normalisation in meta-analyses of thermal responses of biological traits. *PeerJ* 6:e4363. DOI:10.7717/peerj.4363.

Lawton P, Lavalli KL (1995) Postlarval, juvenile, adolescent, and adult ecology. In: Factor JR (ed.) *Biology of the Lobster* Homarus americanus. New York, NY, USA: Academic Press Inc., p. 47-88.

Lin S, Shao L, Hui C, Sandhu HS, Fan T, Zhang L, Li F, Ding Y, Shi P
(2018) The effect of temperature on the developmental rates of
seedling emergence and leaf-unfolding in two dwarf bamboo species.
Trees – Struct Funct 32:751-763

MacKenzie BR (1988) Assessment of temperature effects on
interrelationships between stage durations, mortality, and growth in
laboratory-reared *Homarus americanus* Milne Edwards larvae. *J Exp
Mar Biol Ecol* 116:87-98.

Martin TL, Huey RB (2008) Why "suboptimal" is optimal: Jensen's
inequality and ectotherm thermal preferences. *Am Nat* 173:E102-E118.

McLaren IA (1963) Effects of temperature on growth of zooplankton and
the adaptive value of vertical migration. *J Fish Res Board Can* 20:685-
727.

McLeese DW (1956) Effects of temperature, salinity and oxygen on the
survival of the American lobster. *J Fish Res Board Can* 13:247-272.

Miller RJ (1997) Spatial differences in the productivity of American
lobster in Nova Scotia. *Can J Fish Aquat Sci* 54:1613-1618.

Milne Edwards H (1837) *Histoire Naturelle des Crustacés, Comprenant
l'Anatomie, la Physiologie et la Classification de ces Animaux. Vol. 2*
[*Natural History of Crustaceans, Including Anatomy, Physiology and
Classification of these Animals. Vol. 2*]. Paris, France: Librairie
Encyclopédique de Roret.

O'Connor MI, Bruno JF, Galnes SD, Halpern BS, Lester SE, Kinlan BP,
Weiss JM (2007) Temperature control of larval dispersal and the
implications for marine ecology, evolution, and conservation. *Proc Nat
Acad Sci* 104:1266-1271.

Padmavathi C, Katti G, Sailaja V, Padmakumari AP, Jhansilakshmi V,
Prabhakar M, Prasad YG (2013) Temperature thresholds and thermal
requirements for the development of the rice leaf folder,
Cnaphalocrocis medinalis. *J Insect Sci* 13:96. DOI:10.1673/031.013.
9601.

Pechenik JA (1999) On the advantages and disadvantages of larval stages
in benthic marine invertebrate life cycles. *Mar Ecol Prog Ser* 177:269-
297.

Phillips BF, Sastry AN (1980) Larval ecology. In: Cobb JS, Phillips BF (eds.) *The Biology and Management of Lobsters Volume II: Ecology and Management.* New York, NY, USA: Academic Press, p. 11-57.

Pineda J, Reyns N (2018) Larval transport in the coastal zone: biological and physical processes. In: Carrier TJ, Reitzel AM, Heyland A (eds.) *Evolutionary Ecology of Marine Invertebrate Larvae.* Oxford, UK: Oxford University Press, p. 141-159.

Pinksy ML, Worm B, Fogarty MJ, Samiento JL, Levin SA (2013) Marine taxa track local climate velocities. *Science* 341:1239-1242.

Quinn BK (2017a) Threshold temperatures for performance and survival of American lobster larvae: a review of current knowledge and implications to modeling impacts of climate change. *Fish Res* 186, Part 1:383-396.

Quinn BK (2017b) A critical review of the use and performance of different function types for modeling temperature-dependent development of arthropod larvae. *J Therm Biol* 63:65-77.

Quinn BK (2019) Occurrence and predictive utility of isochronal, equiproportional, and other types of development among arthropods. *Arthropod Struct Dev*, in press. DOI:10.1016/j.asd.2018.11.007.

Quinn BK, Chassé J, Rochette R (2017) Potential connectivity among American lobster fisheries as a result of larval drift across the species' range in eastern North America. *Can J Fish Aquat Sci* 74:1549-1563.

Quinn BK, Rochette R (2015) Potential effect of variation of water temperature on development time of American lobster larvae. *ICES J Mar Sci* 72 (Suppl 1):i79-i90.

Quinn BK, Rochette R, Ouellet P, Sainte-Marie B (2013) Effect of temperature on development rate of larvae from cold-water American lobster (*Homarus americanus*). *J Crustac Biol* 33:527-536. [Corrigendum: *J Crustac Biol* 34:126.]

R Core Team (2014) *R: A Language and Environment for Statistical Computing.* Vienna, Austria: R Foundation for Statistical Computing. https://www.R-project.org/.

Sastry AN, Vargo SL (1977) Variations in the physiological responses of crustacean larvae to temperature. In: Vernberg FJ, Calabrese A,

Thurberg FP, Vernberg WB (eds.) *Physiological Responses of Marine Biota to Pollutants*. New York, NY, USA: Academic Press Inc., p. 401-424.

Schmalenbach I, Franke HD (2010) Potential impact of climate warming on the recruitment of an economically and ecologically important species, the European lobster (*Homarus gammarus*) at Helgoland, North Sea. *Mar Biol* 157:1127-1135.

Schoolfield RM, Sharpe PJH, Magnuson CE (1981) Non-linear regression of biological temperature-dependent rate models based on absolute reaction-rate theory. *J Theor Biol* 88:719-731.

Sharpe PJH, DeMichele DW (1977) Reaction kinetics of poikilotherm development. *J Theor Biol* 64:648-670.

Shi P, Ikemoto T, Egami C, Sun Y, Ge F (2011) A modified program for estimating the parameters of the SSI model. *Environ Entomol* 40:462-469.

Shi P, Li B, Ge F (2012a) Intrinsic optimum temperature of the diamondback moth and its ecological meaning. *Environ Entomol* 41:714-722.

Shi P, Quinn BK, Zhang Y, Bao X, Lin S (2019) Comparison of the intrinsic optimum temperatures for seed germination between two bamboo species based on a thermo-dynamic model. *Global Ecol Conserv* 7:e00568. DOI: 10.1016/j.gecco.2019.e00568.

Shi P, Reddy GVP, Chen L, Ge F (2017) Comparison of thermal performance equations in describing temperature-dependent developmental rates of insects: (II) two thermodynamic models. *Ann Entomol Soc Am* 110:113-120.

Shi P, Sandhu HS, Ge F (2013) Could the intrinsic rate of increase represent the fitness in terrestrial ectotherms? *J Therm Biol* 38:148-151.

Shi P, Wang B, Ayres MP, Ge F, Zhong L, Li B (2012b) Influence of temperature on the northern distribution limits of *Scirpophaga incertulas* Walker (Lepidoptera: Pyralidae) in China. *J Therm Biol* 37:130-137.

Sreedevi G, Prasad YG, Prabhakar M, Rao GR, Vennila S, Venkateswarlu B (2013) Bioclimatic thresholds, thermal constants and survival of mealybug, *Phenacoccus solenopsis* (Hemiptera: Pseudococcidae) in response to constant temperatures on hibiscus. *PLoS ONE* 8:e75636. DOI:10.1371/journal.pone.0075636.

Templeman W (1936) The influence of temperature, salinity, light and food conditions on the survival and growth of the larvae of the lobster (*Homarus americanus*). *J Biol Board Can* 2:485-497.

Van Olst JC, Ford RF, Carlberg JM, Dorband WR (1976) Use of thermal effluent in culturing the American lobster. In: *Power Plant Waste Heat Utilization in Aquaculture, Workshop I.* Newark, NJ, USA: Public Service Electric & Gas Company, p. 71-100.

Vaughn D, Allen JD (2010) The peril of the plankton. *Integr Comp Biol* 50:552-570.

Wahle RA, Castro KM, Tully O, Cobb JS (2013) *Homarus.* In: Phillips BF (ed.) *Lobsters: Biology, Management, Aquaculture and Fisheries* (2nd Ed.). Oxford, UK: John Wiley and Sons Ltd., p. 221-258.

Wahle RA, Incze LS, Fogarty MJ (2004) First projections of American lobster fishery recruitment using a settlement index and variable growth. *Bull Mar Sci* 74:101-114.

Waller JD, Wahle RA, McVeigh H, Fields DM (2017) Linking rising pCO_2 and temperatures to the larval development and physiology of the American lobster (*Homarus americanus*). *ICES J Mar Sci* 74:1210-1219.

Winberg GG (1971) *Methods for the Estimation of Production of Aquatic Animals.* London, UK: Academic Press Inc., 175 p.

Yamamoto T, Jinbo T, Hamasaki K (2017) Intrinsic optimum temperature for the development of decapod crustacean larvae based on a thermodynamic model. *J Crustac Biol* 37:272-277.

Yao N, Zhang Y (2018) Investigating temporal variation in stable carbon and nitrogen isotope values of Florida Caribbean spiny lobster (*Panulirus argus*) recruits. *Bull Mar Sci* 94:847-861.

In: Lobsters
Editor: Brady K. Quinn

ISBN: 978-1-53615-711-6
© 2019 Nova Science Publishers, Inc.

Chapter 2

THE CRYPTIC *HOMARUS GAMMARUS* (L., 1758) JUVENILES: A COMPARATIVE APPROACH TO THE MYSTERY OF THEIR WHEREABOUTS

Gro I. van der Meeren[1,] and Astrid K. Woll[2]*
[1]Institute of Marine Research, NO-5817 Bergen, Norway
[2]Woll Naturfoto, Industrivegen 9, NO-6475 Midsund, Norway

ABSTRACT

A series of selected biological, ecological, and morphological traits of North Atlantic decapod crustaceans were used to discuss the possible biotope of juvenile European lobsters, *Homarus gammarus* (Linnaeus, 1758), and how to approach filling our knowledge gaps regarding the whereabouts of this life stage in nature. The information examined provided a review of what is known and what is anticipated regarding this species. This comparative and holistic ecological approach will provide

* Corresponding Author's E-mail: GroM@hi.no.

useful information for future studies of the 'missing link' in this species' life history – small, juvenile European lobsters. Three hypotheses for the reasons these life stage have not been found are that they are: (1) located too deep to be found in shoreline-based field studies; (2) distributed too scarcely to be found; or (3) living deep inside shelters, which are unapproachable by competitors, predators, and humans. We compared the ecomorphology, life history, habitat, and behavior of *H. gammarus* to those of *Homarus americanus* H. Milne Edwards, 1837, as well as to those of other decapod species living within a similar geographical range to that of adult lobsters. The main differences found among the biotopes of the homarid lobsters were in the diversity of their predators and the number of competitive decapods present, which are both higher in European shallow-water cobble bottom areas. This should have strong impacts on the survival probabilities of settling lobsters. Larvae and adults of *H. gammarus* are found at much lower densities than those of *H. americanus*, although early-stage larvae of both species are regularly caught in light traps and plankton nets. Field and laboratory studies indicate limited dispersal of larvae in *H. gammarus*. In the laboratory, settling juveniles of both species show clear preferences for shelters by burrowing underneath or beside solid objects, which is symptomatic of a cryptic life habit. Juvenile *H. gammarus* show traits that are suggestive of adaptations to shallow habitats, even potentially to intertidal life. The higher biodiversity in European shallow waters may cause the young lobsters to stay sheltered deeper within the substrate in shallow places that competitors and predators cannot reach.

Keywords: American lobster, biology, decapod crustaceans, ecology, European lobster, life history, morphology, North Atlantic Ocean

1. INTRODUCTION

Traditionally, the acquisition of information on wild-living species used to be focused on assessing the presence and life habits of mature individuals. However, in the last century the importance of learning about the full life history to understand the ecology of an organism has become apparent, and is now known to be crucial to protecting and managing wild populations sustainably (Sadovy 2001; Denney et al. 2002). Lobsters, particularly the homarid species (*Homarus* spp.), are potentially vulnerable to overexploitation because they are long-lived and of low to medium fecundity (Agnalt et al. 1999; Wahle et al. 2013a). However, the whereabouts of the early life stages of the European lobster, *Homarus gammarus* (Linnaeus, 1758), are still an unsolved mystery (Mercer et al. 2001; van der Meeren 2005).

While little is known about the larger, commercially important decapod crustacean species of the Northern Atlantic Ocean, there are few species for which there are such limited field observations of critical life history stages as for *H. gammarus*. However, it is possible that more could be learned about these concealed life stages by comparing this lobster's living biotope, morphology, and behavior with those of other decapod crustaceans. Comparing behavioral traits is an established method in evolutionary science, which has been used to define phylogenetic relationships and evolutionary connections among taxa (Clutton-Brock and Harvey 1984; Harvey and Nee 1997). Ecomorphological adaptation has also been used as a tool to look at evolutionary habitat shifts in crabs (*Cancer* spp.) by analyzing such traits as size and habitat (Harrison and Crespi 1999). Life history traits include parental investment, which is closely connected to the ecological constraints placed on parents and offspring (Clutton-Brock and Godfray 1991). A comparison of such traits among a selection of North Atlantic decapod crustacean species was presented and discussed in this chapter. The reasoning behind this approach was that evolution and adaptation should have resulted in each species' body shape, pigmentation, environmental requirements, settlement and establishment choices, and behaviors in general being adapted to their optimal living biotope.

This chapter investigated and compared a selection of North Atlantic decapod crustacean species (Figure 1). This included: three nephropid lobsters, the European lobster *H. gammarus*, American lobster *Homarus americanus* (H. Milne Edwards, 1837), and Norway lobster *Nephrops norvegicus* (L., 1758), as well as the limnic noble crayfish *Astacus astacus* (L., 1758); brachyuran crabs, including the brown or edible crab *Cancer pagurus* (L., 1758) and green shore crab *Carcinus maenas* (L., 1758); the European spiny lobster *Palinurus elephas* (Fabricius, 1787); anomurans, such as the spiny squat lobster *Galathea strigosa* (L., 1761) and red king crab *Paralithodes camtschaticus* (Tilesius, 1815); and natantian species, including the hooded shrimp *Athanas nitescens* (Leach, 1813) and northern deep-water shrimp *Pandalus borealis* Krøyer, 1838.

The above species are all temperate- to cold-water-adapted and occupy the same trophic level spectrum (2 to 3), but live at different depths, have different habitats and diurnal habits, and face different levels of competition and predation pressure (Williamson 1905; Pearson 1908; Herrick 1911; Crothers 1967, 1968; Christiansen 1969; Ansell and Robb 1977; Cooper and Uzmann 1980; Shumway et al. 1985; Factor 1995; Hunter 1999; Bergström 2000; Ulmestrand and Eggert 2001; Goñi and Latrouite 2005; Poore et al. 2011; Groeneveld et al. 2013; Stevens 2014; Sundet 2014; Füreder 2016). The little information found on *A. nitescens* was limited to the traits most of the other species have in common (Palomares and Pauly 2018), so this species was not included in the statistical comparisons performed. Even when information had been published on the traits of mature individuals of the rest of these species, information on the juvenile stages is lacking for several of them, not only *H. gammarus*. Some of the information that is available comes from laboratory studies, which may not extrapolate exactly to wild individuals. Nevertheless, our approach is an attempt that may possibly deduce the potential living areas and ecology of the hitherto cryptic life stages of young *H. gammarus*.

2. METHODS

Interspecific similarities and differences in traits among species and their implications were assembled and analyzed for the following categories and types of traits:

- Morphology (Table 1): body shape; shape and relative strength of pereiopods, including chelae; spines in the larval stages; abdominal (tail) flexibility; carapace spines (adults and juveniles);

length of second antennae relative to body length; and pigmentation.

- Life history (Table 2): Maximum age (lifespan); age of maturation; maximum size; fecundity; duration of larval period; and size at settling.
- Biotope (Table 3): use of shelter; habitat type; depth; competitors (other decapod species and groups); and predator diversity.
- Behavior (Table 4): Burrowing; tail flip escape response; nocturnal activity; level of aggression; social behavior; and diet.
- Other parameters (Table 5): Temperature range; and trophic level.

The above traits were analyzed and classified using cladistics and principal component analysis (PCA), inspired by the methods used for natural classification by Nelson and Platnick (1981) and Williams and Ebach (2004). Results of statistical analyses were discussed and related to the specific traits of each species, in both its mature and juvenile life stages. The significances of particular traits in relation to life history strategies were also discussed. The traits compared are shown in Tables 1-5.

For analyses, the phenotypic trait categories from Tables 1-5 were represented as numerical variables prior to performing clustering analyses (Tables 6 and 7). Clustering analyses were fitted by the 'Agnes' method (Kaufman and Rousseeuw 1990), implemented in the '*cluster*' package (Maechler et al. 2018) in R v.3.5.1 (R Core Team 2016).

The effects of ecologically important traits (predation pressure and competitive species richness) for the two *Homarus* species were analyzed to look at their possible impacts on survival strategies used in the juvenile phase of these two species.

Table 1. The selected morphological traits (body shape, shape and relative strength of pereiopods, including chelae, spines in the larval stages, abdominal (tail) flexibility, presence of carapace spines (adults and juveniles), length of second antennae relative to body length, and pigmentation) compared among adults and juveniles of the selected decapod species

Species	Body shape	Pereiopods/ claw shape	Chelae shape	Chelae	Larval spines	Flexible tail	Carapace spines	Relative 2nd antennae length	Pigmentation
Adults									
Homarus gammarus (Hg)	Elongate	Strong, pincers	Pointed	Strong	No	Yes	Small	> body	Dark
Homarus americanus (Ha)	Elongate	Strong, pincers	Pointed	Strong	No	Yes	Small	> body	Medium
Paralithodes camtschaticus (Pc)	Wide	Strong, pointed	Pointed	Strong	No	No	None	< carapace	Reddish
Cancer pagurus (Cp)	Wide	Long, strong, pointed	Pointed	Weak	No	No	None	< carapace	Brown
Carcinus maenas (Cm)	Wide	Strong, pointed	Pointed	Strong	No	No	None	< carapace	Greenish
Pandalus borealis (Pb)	Elongate	Long, slender,	Pointed	Weak	Yes	Yes	Yes	> body	Reddish
Palinurus elephas (Pe)	Elongate	Strong	Hooked	None	Yes	Yes	Yes	> body	Contrasting patterns
Nephrops norvegicus (Nn)	Elongate	Slender, pincers	Pointed	Slender	No	Yes	Small	> body	Reddish
Galathea strigosa (Gs)	Elongate	Strong	Hooked	Weak	No	Yes	Small	> carapace	Contrasting patterns
Astacus astacus (Aa)	Elongate	Strong	Pointes	Strong	No	Yes	None	> carapace	Dark
Juveniles, settlers									
Hg juv	Elongate	Thin	Pointed	Slender	No	Yes	Small	> body	Brownish
Ha juv	Elongate	Thin	Pointed	Slender	No	Yes	Small	> body	Dark

Species	Body shape	Pereiopods/ claw shape	Chelae shape	Chelae	Larval spines	Flexible tail	Carapace spines	Relative 2nd antennae length	Pigmentation
Pc juv	Wide	Strong	Pointed	Strong	No	No	Small	< carapace	Reddish
Cp juv	Wide	Long, strong	Pointed	Weak	Yes	No	None	< carapace	Pale
Cm juv	Wide	Strong	Pointed	Weak	No	No	None	< carapace	Patterns
Pb juv	Elongate	Thin	Pointed	Yes	Yes	Yes	Yes	> body	Reddish
Pe juv	Elongate	Thin	Hooked	No	Yes	Yes	Small	> body	Pale
Nn juv	Elongate	Thin,	Pointed	Slender	No	Yes	None	> body	Reddish
Gs juv	Elongate	Thin	Hooked	Weak	Yes	Yes	Yes	> carapace	Pale
Aa juv	Elongate	Strong	Pointed	Strong	No	Yes	None	> carapace	Dark

References: Pearson 1908; Herrick 1911; Crothers 1967; 1968; Christiansen 1969; Ansell and Robb 1977; Cooper and Uzmann 1980; Shumway et al. 1985; Factor 1995; Fogarty 1995; Hunter 1999; Sheehy et al. 1999; Bergström 2000; Ulmestrand and Eggert 2001; Goñi and Latrouite 2005; Agnalt 2008; Poore et al. 2011; Groeneveld et al. 2013; Stevens 2014; Sundet 2014; Füreder 2016; MarLIN 2016; Palomares and Pauly 2018.

Table 2. The selected life history traits (maximum age/lifespan, age of maturation, maximum size, fecundity, duration of larval period, and size at settling) compared among adults and juveniles of the selected decapod species

Species	Lifespan (conservative estimate)	Maturation age, females	Max size (mm CL or CW)	Fecundity	Larval period (northern distribution)	Size at settling (mm CL or CW)
Adults						
Homarus gammarus (Hg)	50	7	160	Medium	NA	NA
Homarus americanus (Ha)	50	5	220	High	NA	NA
Paralithodes camtschaticus (Pc)	15	4	220	High	NA	NA
Cancer pagurus (Cp)	15	3	160	High	NA	NA
Carcinus maenas (Cm)	8	3	70	High	NA	NA
Pandalus borealis (Pb)	6	4	35	Low	NA	NA
Palinurus elephas (Pe)	25	5	150	High	NA	NA
Nephrops norvegicus (Nn)	10	4	50	Medium	NA	NA
Galathea strigosa (Gs)	?	?	53	Medium	NA	NA
Astacus astacus (Aa)	15	3	120	Low	NA	NA
Juveniles, early settlers						
Hg juv	NA	NA	NA	NA	Weeks	10
Ha juv	NA	NA	NA	NA	Weeks	12
Pc juv	NA	NA	NA	NA	Weeks	1
Cp juv	NA	NA	NA	NA	Months	5
Cm juv	NA	NA	NA	NA	Months	1
Pb juv	NA	NA	NA	NA	Weeks	3
Pe juv	NA	NA	NA	NA	Months	25
Nn juv	NA	NA	NA	NA	Months	7
Gs juv	NA	NA	NA	NA	Weeks	1
Aa juv	NA	NA	NA	NA	No free larval period	3

References: See Table 1.

The PCA conducted reduced N descriptive variables to n independent principal components (PCs). Retaining the PCs that explained the greatest proportions of the variance in the data reduced the dimensionality of the data to allow for greater ease of interpretation of the patterns present.

Figure 1. Selected decapod species with home ranges in the Northern Atlantic region of Europe, from upper-left to lower-right: *Homarus gammarus* (Linnaeus, 1758), *Homarus americanus* H Milne Edwards, 1837, *Nephrops norvegicus* (Linnaeus, 1758), *Pandalus borealis* Krøyer, 1838, *Galathea strigosa*, (Linnaeus, 1761), *Athanas nitescens* (Leach, 1813), *Cancer pagurus* (Linnaeus, 1758), *Paralithodes camtschaticus* (Tilesius, 1815), *Carcinus maenas* (Linnaeus, 1758). Two non-native species, *Homarus americanus* and *Paralithodes camtschaticus*, are present in northeastern Atlantic and Barents Sea waters. 'Lobster-shaped' species in upper section, 'crab-shaped' decapods in the lower section. Photographers: *P. borealis* by Ø. Paulsen/IMR, *A. nitescens* by Nils Aukan, the rest by AK Woll.

Table 3. The selected biotope traits (use of shelter, habitat type, depth, competitors (other decapod species and groups), and predator diversity) compared among adults and juveniles of the selected decapod species

Species	Shelter	Habitat	Depth (m)	Decapod competitor diversity	Predator diversity
Adults					
Homarus gammarus (Hg)	Burrows under solid objects	Mixed	0 - 50	High	High
Homarus americanus (Ha)	Burrows	Rocky and soft	0 - 100	Low	Medium
Paralithodes camtschaticus (Pc)	In pods (high density spots)	Mixed	0 - 300	Low	Medium
Cancer pagurus (Cp)	In cracks and crevices, in sand	Rocky and soft bottoms	0 - 70	High	High
Carcinus maenas (Cm)	In cracks and crevices, in sand	Rocky	0 - 10	High	High
Pandalus borealis (Pb)	High density	Soft bottoms	9 - 1500	Low	High
Palinurus elephas (Pe)	Under solid objects, in crevices	Rocky bottoms	0 - 200	Medium	High
Nephrops norvegicus (Nn)	Deep burrows	Muddy sediments	20 - 800	Low	High
Galathea strigosa (Gs)	In cracks and crevices	Rocky bottoms	0 - 10	Medium	High
Astacus astacus (Aa)	Burrows, under objects	Mixed mud/rocks	0 - 15	Low	Low
Juveniles, early settlers					
Hg juv	Burrows/cracks	Mixed	0 - 10	High	High
Ha juv	In cracks and crevices	Rocky	0 -5	Medium	Medium
Pc juv	Habitat-building organisms	Rocky, kelp forests	0 - 5	Medium	Medium
Cp juv	On sand	Mostly rocky bottoms	0 - 10	High	High
Cm juv	Between sand grains	Sandy	0 - 2	High	High
Pb juv	None	Sandy	> 50 m	Medium	High
Pe juv	Mussels/crevices	Rocky	10 - 90	Medium	High
Nn juv	In burrows	Muddy	20 - 300	Low	High
Gs juv	In cracks and crevices	Rocky	0 - 10	Medium	High
Aa juv	Burrows, under objects	Mixed mud/rocks	0 - 15	Low	Low

References: See Table 1.

Table 4. The selected behavioral traits (burrowing, tail flip escape response, nocturnal activity, level of aggression, social behavior, and diet) compared among adults and juveniles of the selected decapod species

Species	Active burrower	Tail flip escape	Strictly nocturnal	Aggression	Social	Seasonal migration	Diet
Adults							
Homarus gammarus (Hg)	Yes	Yes	Yes	Yes	No	Yes	Omnivore
Homarus americanus (Ha)	Yes	Yes	Yes	Yes	No	Yes	Omnivore
Paralithodes camtschaticus (Pc)	Occasional	No	No	No	Yes	Yes	Omnivore
Cancer pagurus (Cp)	Occasional	No	No	Medium	Medium	Yes	Omnivore
Carcinus maenas (Cm)	Occasional	No	No	Medium	Medium	No	Omnivore
Pandalus borealis (Pb)	No	Yes	No	Low	Yes	No	Planktivore
Palinurus elephas (Pe)	No	Yes	Yes	Low	Yes	Yes	Omnivore
Nephrops norvegicus (Nn)	Yes	Yes	Yes	High	No	No	Omnivore
Galathea strigosa (Gs)	No	Yes	Yes	Low	Medium	No	Omnivore
Astacus astacus (Aa)	No	yes	Yes	Yes	Medium	No	Omnivore
Juveniles, early settlers							
Hg juv	Yes	Yes	Yes	Yes	No	No	Omnivore
Ha juv	Yes	Yes	Yes	High	No	No	Omnivore
Pc juv	No	No	Yes	Low	Yes	No	Omnivore
Cp juv	No	No	Yes	Medium	No	No	Omnivore
Cm juv	Yes	No	Yes	Low	No	No	Omnivore
Pb juv	No	Yes	No	Low	Yes	No	Omnivore
Pe juv	No	Yes	Yes	Low	No	No	Omnivore
Nn juv	Yes	Yes	Yes	Yes	No	No	Omnivore
Gs juv	No	Yes	Yes	Low	Medium	No	Omnivore
Aa juv	No	Yes	Yes	Medium	Medium	No	Omnivore

References: See Table 1.

Table 5. Other traits (temperature range and trophic level) compared among adults and juveniles of the selected decapod species

Species	Temperature range (°C)	Trophic level
Adults		
Homarus gammarus (Hg)	10 - 22	2
Homarus americanus (Ha)	5 - 25	3
Paralithodes camtschaticus (Pc)	- 1 - 11	2
Cancer pagurus (Cp)	9 - 14	2
Carcinus maenas (Cm)	10 - 25	2
Pandalus borealis (Pb)	0 - 5	2
Palinurus elephas (Pe)	12 - 16	2
Nephrops norvegicus (Nn)	5 - 10	3
Galathea strigosa (Gs)	5 - 20	2
Astacus astacus (Aa)	0 - 18	2
Juveniles, early settlers		
Hg juv	15 - 20	2
Ha juv	12 - 20	2
Pc juv	0 - 3	2
Cp juv	10 - 15	2
Cm juv	10 - 15	2
Pb juv	0 - 5	2
Pe juv	15 - 20	2
Nn juv	8 - 18	2
Gs juv	5 - 20	2
Aa juv	0 - 18	2

References: See Table 1.

In the PCA, the first three and six principal axes explained 59% and 84% of the total variance, respectively. Subsequent analyses were thus based on the first six out of 19 principal component axes.

The traits from all sections were included in the main analyses, but the significance of the different types of traits were discussed in separate sections when comparing species and life stages.

The trait information analyzed was collected from the trait databases at SeaLifeBase.org (Palomares and Pauly 2018) and MarLIN.uk.ac (MarLIN 2016), from scientific papers and books (cited in Tables 1-5), and, for some traits, the authors' personal observations and observations by experienced lobster professionals.

Table 6. The phenotypic trait categories in Tables 1-5 translated into numerical variables prior to fitting in clustering analyses for each and all categories for adults

Trait/code	1	2	3	4
Morphological traits				
Body shape	Elongate	Wide	NA	NA
Pereiopods	Slender	Long	Strong	NA
Pereiopod claw shape	Pointed	Hook	Paddle-like	NA
Chelae	None	Scooped	Pointed	Strong, rounded
Secondary chelae	No	Yes	NA	NA
Rostrum	No	Short	Prolonged	NA
Flexible tail	No	Yes		
Spines on carapace	No	Small	Prominent	NA
Relative 2nd antennae length	> body length	= body	< carapace	NA
Pigmentation	Dark	Medium, brownish	Pale, reddish	NA
Life history traits				
Lifespan (estimated in years)	< 3	3 - 15	> 15	NA
Maturation age, females (years)	< 2	2 - 5	> 5	NA
Max size (mm CL or CW)	< 55	55 - 200	> 200	NA
Fecundity	< 50 000	50 000	> 50 000	NA
Habitat traits				
Shelter	None	Burrow	Cracks	Pods
Habitat	Soft	Algae	Rocky, mixed	Hard
Depth (m)	0 - 10	0 - 50	> 50	NA
Behavioral traits				
Active burrower	No	Occasionally	Yes	NA
Tail flip escape	No	Yes	NA	NA
Strictly nocturnal	No	Yes	NA	NA
Interspecific aggression	No	Medium	High	NA
Social	No	Medium	High	NA
Seasonal migration (females)	No	Yes		
Diet	Planktivore	Omnivorous	Predatory	NA
Local decapod competitor diversity	Low (< 5 species)	Medium (5-10 species)	High (> 10 species)	NA
Predator diversity	Low (< 5 species)	Medium (5-10 species)	High (> 10 species)	NA
Other parameters				
Maximum temperature (°C)	< 5	5 - 15	15 - 25	NA
Trait/code	1	2	3	4
Optimal temperature (°C)	< 5	5 - 10	10 - 20	NA
Trophic level	< 2	2 - 3.9	> 4	NA

References: See Table 1.

Table 7. The phenotypic trait categories in Tables 1-5 translated into numerical variables prior to fitting in clustering analyses for each and all categories for juveniles

Trait/code	1	2	3
Morphological traits			
Body shape	Elongate	Wide	NA
Pereiopods	Strong	Long	Thin
Pereiopod claw shape	Pointed	Hook	NA
Chelae	None	Weak	Strong
Secondary chelae	None	Yes	NA
Rostrum	None	Short	Prolonged
Flexible tail	None	Yes	NA
Spines on carapace	None	Spread	Dense
Relative 2nd antennae length	> body length	= body	< carapace
Pigmenation	Dark	Medium, brownish	Pale, reddish
Trait: Life history			
Lifespan (estimated, in years)	< 2	2-5	> 5
Size at settling (mm CL or CW)	< 1	1-9	> 10
Biotope traits			
Shelter	None	Burrow	Cracks
Habitat	Soft	Hard	Mixed
Depth (m)	< 0.5	0.5-10	> 10
Behavioral traits			
Active burrower	No	Occasionally	Yes
Tail flip escape	No	Yes	NA
Strictly nocturnal	No	Yes	NA
Interspecific aggression	No	Medium	High
Social	No	Medium	High
Seasonal migration (females)	No	Yes	
Diet	Vegetarian	Omnivorous	Predator
Local decapod competitor diversity	Low (< 5 species)	Medium (5-10 species)	High (> 10 species)
Predator diversity	Low (< 5 species)	Medium (5-10 species)	High (> 10 species)
Other parameters			
Maximum temperature (°C)	< 5	5 - 15	15 - 25
Optimal temperature (°C)	< 5	5 - 10	10 - 20
Trait/code	1	2	3
Trophic level	< 2	2 - 3.9	> 4

References: See Table 1.

3. RESULTS

3.1. Overall Comparisons among Decapods

The results of the main clustering analysis are presented in Figure 2. On the dendrogram, the adult and juvenile stages of four species clustered together as each other's nearest neighbors (*A. astacus*, *P. borealis*, *N. norvegicus*, and *P. camtschaticus*). For the remaining species, the nearest neighbor of a group was usually the same life stage of another species.

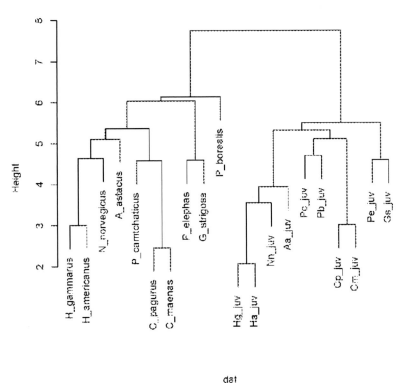

Figure 2. Clustering of the species, as adults and juveniles ('juv'), by the 'Agnes' method, based on data for all the traits listed in Tables 1-5. Agglomerative coefficient: 0.51. Statistics were performed by Francois Besnier, IMR.

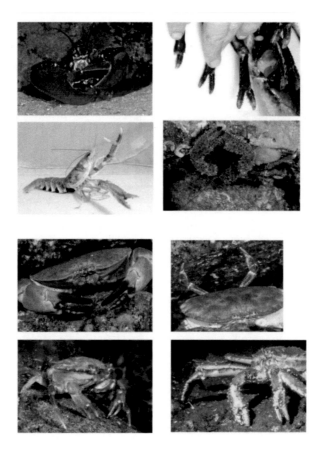

Figure 3. Variation in morphological details between several species, with emphasis on chelae and claws and pereiopods, from upper-left to lower-right: *Homarus gammarus* (L., 1758): chelae, and clawed pereiopods, *Nephrops norvegicus* (L., 1758) chelae and weak pereiopods, *Galathea strigosa* (L., 1761): chelae and hooked claws, *Cancer pagurus* (L., 1758) chelae (left) and straight claws on pereiopods (right), *Carcinus maenas* (L., 1758) chelae and claws, *Paralithodes camtschaticus* (Tilesius, 1815) chelae and claws. Photographer: AK Woll.

In such cases, the two species formed two clusters, where the first cluster consisted of the two adult stages and the second consisted of the two juvenile life stages; for example, this was found for *H. gammarus* and *H. americanus*. For these species, it can be inferred that the amount of dissimilarities between species at the same life stage was less than the amount of dissimilarities within the same species between different life stages.

However, for *H. gammarus*, the information on the life habit and biotope of juveniles (< 25 mm carapace length (CL)) included in the above analyses was based on laboratory studies. This was because these juveniles have never been seen in the field, except for a handful of single occasions in different habitats over a 40-year period. Anecdotal observations were made, but not published, of one *H. gammarus* juvenile in an empty mussel shell in shallow water in Sweden (observed by B. Dybern in the 1970s), one in a pearl net for oysters in Denmark (observed by C. Formsgaard Nielsen in 2015), and one in a crack between rocks in Norway (observed by A. K. Woll in 2015, see Figure 4). However, in the laboratory the two species behave very similarly (Wahle et al. 2013a), as is also reflected in the limited juvenile data available in Tables 1-5. The nephropid lobsters clustered together, and separated into one juvenile and one adult cluster (Figure 2), but the juvenile *H. gammarus* was more distinct from the juvenile *H. Americanus* than was expected from the similarities between the adult lobsters. Therefore, although information on wild juveniles is lacking, it seems that based on laboratory observations and inferred ecological traits there are several important differences between juvenile *H. gammarus* and *H. americanus*.

3.2. Morphology

If bodily structure and function are closely connected to the ecological niche of a species (Dullemeijr 1974; Alexander 1988; Bock 1994), then it may be feasible to detect common ecological groups in decapods by comparing their morphological traits (e.g., body shape, sense organs, and the function of the tail, chelae, and pereiopods) by looking at their ecomorphology, as well as their functional morphology. The 'Agnes' analyses clustered shrimps, crayfish and lobsters, crabs, including *P camtschaticus* and galatheans, and *P. elephas* into three distinct clusters (agglomerate coefficient = 0.68). For juveniles, the functionality of body shapes and appendages is closely related to their chances of survival. The crab-shaped decapods *C. pagurus*, *C. maenas*, and *P. camtschaticus*, with wide carapaces, also all seem to have relatively short antennae and an abdomen that is no more than a reduced flap tucked underneath the body. The long, slender *P. borealis*, *A. astacus*, *P. elephas*, *H. americanus*, and *H. gammarus* (and, to some extent, the prolonged, although still quite wide, *G. strigosa*) all have long to very long second antennae, and all have strong, flexible abdomens (see Figure 3).

From these few morphological traits, it can be deduced that crab-like decapods could have different feeding and survival strategies from those of the more slender (shrimp- or lobster-like) species. The survival strategies of crab-like species appear to involve growing to a larger size (*P. camthschaticus* and *C. pagurus*) and having a more solid carapace with no vulnerable abdomen (*C. pagurus* and *C. maenas*). Conversely, the more slender species are better adapted for speed and making rapid escapes, although the longer, jointed tail they use for these purposes may be more vulnerable to attacks in open spaces.

Crabs and spiny lobsters, which live in the same region and partly in the same biotopes, are usually lighter in color, ranging from brown to spotted, some with starkly contrasting pigmentation providing color-camouflage in the surrounding habitat. They also tend to change pigmentation from lighter and spotted to more homogenous in the first years of life. All the deep-living species examined are reddish in color, from the brightly red *P. borealis* to the pale red *N. norvegicus* and light brownish red king crab. These species do not change their pigmentation as they grow. Red light does not penetrate very far into water, so red colors provide good camouflage in deeper waters. Pigmentation ranges from black through dark blue to dark brown or grayish brown in *H. gammarus* and *A. astacus*. *H. gammarus* tend to be all black in the northern part of their range, usually with some white markings near the mouth, while the coloration of *H. americanus* is more uniform throughout its range, with dark green and reddish brown undertones and less pronounced pale spots in the head region (Figure 3). Both lobster species seem to develop dark pigmentation at settlement, although this probably depends on food quality. Laboratory-reared juvenile lobsters tend to be paler than wild-caught juveniles if they are fed an astaxanthin-deficient diet (D'Agostino 1980; D'Abramo et al. 1983; Lim et al. 1997; Tlusty and Hyland 2005). Juvenile *H. gammarus* also respond with marked changes in pigmentation when their feed is changed by adding astaxanthin or by swithching it from *Artemia* to mussels in the laboratory (Wickins and Lee 2002; Drengstig et al. 2003; Kristiansen et al. 2004). The dark pigmentation of lobsters may provide better camouflage in deep cracks and within caverns than the brown and green coloration of crabs. Lobsters are quite conspicuous on open ground, however, so their dark pigmentation can be seen as being linked to a cryptic lifestyle, in which they aim to stay hidden from predators on the sea bottom.

Figure 4. Newly spawned egg clutches and early life stages, listed from upper-left corner to lower-right. Egg clutches: *Homarus gammarus* (L., 1758), *Nephrops norvegicus* (L., 1758), *Cancer pagurus* (L., 1758), *Carcinus maenas* (L., 1758), *Paralithodes camtschaticus* (Tilesius, 1815). Juveniles: *H. gammarus* (stage V, reared) and a small juvenile found in a crevice January, May and August 2016, *C. pagurus*, *C. maenas*, *P. camtschaticus*. Photographers: Egg clutch *H. Gammarus* by Finn Refsnes; Juvenile *H. gammarus* (reared) by GI van der Meeren/IMR and *C. maenas* by IMR, the rest by AK Woll.

3.3. Life History

The 'Agnes' analyses of life history traits showed that the two *Homarus* species were closely clustered together, and seemed to have traits in common as both adults (agglomerate coefficient = 0.58) and juveniles (agglomerate coefficient = 0.64).

Internal mating, wherein the male deposit a sperm package in the spermathecae of the female, is a common feature of all decapods (Subramoniam 2016). However, besides this, the selection of North Atlantic species examined has a wide range of life history strategies, including very different cycles of reproduction and recruitment. Indeed, fecundity was very dissimilar among these species, ranging from low in *A. astacus* to high in *C. pagurus* (Pearson 1908; Herrick 1911; Cooper and Uzmann 1980; Shumway et al. 1985; Shields 1991; Ingle 1992; Factor 1995; Hunter 1999; Bergström 2000; Poore et al. 2011; Groeneveld et al. 2013; Stevens 2014; Füreder 2016; Bakke et al. 2018) (Figure 4).

Larval development is also widely different among these species, from the larval stages all being completed before hatching and extended parent care being provided to the newly molted juveniles in *A. astacus* to the prolonged larval stages of a year or more in duration in *P. elephas*. All the marine species examined release planktonic larvae at hatch, which then develop through several molts before they metamorphose into the settling stage that then molts into the bottom-dwelling juvenile stage. The fecundity of the different species reflects the results of trade-offs between the costs of parental care versus the survival of the brood (Sastry 1983; Clutton-Brock and Godfray 1991; Denney et al. 2002). The size of the eggs depends on how much energy is stored in each egg, and this ranged from the nearly microscopic eggs of the brachyuran crabs to the slightly larger eggs of *P. camtschaticus*, to the relatively large eggs, which hatch into larger larvae, in *P. borealis*, *G. strigosa*, *P. elephas*, and the three nephropid lobster species (Williamson 1983; Ingle 1992). The number of larval molts and time to maturation also vary. *Palinurus elephas* has an extremely long larval period, but homarid lobsters undergo only four larval molts, and may be ready to settle in less than three weeks if the temperature is optimal. Growth from the juvenile stages to maturity seems to be related more to the maximum body size of a species than to its clutch and egg sizes (Tables 2 and 3). Size at settling mostly reflects egg size, as for example newly settled *C. pagurus*, which have larger eggs, have carapace widths (CW) of 2-3 mm, while in *C. maenas*, which has smaller eggs, the carapace widths of new settlers are less than 0.5 mm. The first juvenile stage in both homarid species has a CL of almost 10 mm. However, the less fecund *H. gammarus*, with larger eggs, has slightly larger newly hatched larvae.

Figure 5. Some of the common and efficient visual predators found in shallow European waters (from upper-left to lower-right): Atlantic cod *Gadus morhua* (L., 1758), ballan wrasse *Labrus bergylta* (Ascanius, 1767), goldsinny *Ctenolabrus rupestris* (L. 1758), plaice *Pleuronectes platessa* (L., 1758), curled octopus *Eledone cirrhosa* (Lamarck, 1798), shorthorn sculpin *Myoxocephalus scorpius* (L., 1758). Photographer: AK Woll.

Age is difficult to determine in decapod crustaceans, as reviewed by Vogt (2012). The maximum age of a species seems to be connected with its general size, with larger species generally living longer, and homarid species continue to grow and produce new batches of recruits for over 50 years. However, *H. gammarus* differs from *H. americanus* in the rates of development during the larval stages. *H. gammarus* larvae are larger at hatch, and they start metamorphosis at stage III, becoming more developmentally advanced sooner than those of *H. americanus* and leading to less change between stages III and IV in *H. gammarus* (Rötzer and Haug 2015). After settling, however, juveniles of *H. americanus* grow faster than *H. gammarus*, and individuals of this species can thus reach larger sizes and weights over their lifetime than *H. gammarus* (Lawton and Lavalli 1995). For the other species examined, the lifespan is limited, and for *C. maenas*, which undergoes a terminal molt at an age of six to eight years, possibilities for further growth are terminated and reproduction is limited to one to two years more after the terminal molt (Crothers 1967, 1968).

The underlying causes for the evolution of dissimilarities in fecundity, larval and settling juvenile sizes, and growth potential between the two homarid species are so far not known, but may be discussed in relation to the different ecosystems wherein these species have evolved (see subsequent sections).

3.4. Biotope

In statistical analyses of the species in relation to their biotopes, a new clustering pattern emerged, which was mainly related to the species' habitat depth, substrate preference, predators, and competition (agglomerated coefficient = 0.59). The habitat preferences of the selection of decapods examined range from deep, soft sea bottoms down to several hundred meters to shallow, intertidal cobble grounds (Williamson 1905; Pearson 1908; Herrick 1911; Crothers 1967; Christiansen 1969; Cooper and Uzmann 1980; Shumway et al. 1985; Factor 1995; Hunter 1999; Poore et al. 2011; Groeneveld et al. 2013; Stevens 2014; Füreder 2016). The habitats preferred by mature lobsters of *Homarus* spp. are rocky to mixed bottoms, in relatively shallow waters, particularly in the summer season, where they excavate shelters underneath protective rocks or other hard objects (Wahle et al. 2013b). Although *G. strigosa*, *P. elephas*, and *C. pagurus* are often found in similar habitats, *P. elephas* is mostly distributed in the southern part of the range of *H. gammarus*, while *C. pagurus* may be found in more exposed waters. *G. strigosa* prefers more sheltered and subtidal sites. *C. maenas* and *P. camtschaticus* can be found on both soft and rocky bottoms, while *P. camtschaticus*, *P. borealis*, and *N. norvegicus* are all found in deeper waters with soft or muddy bottoms, on open ground. High-density patches of *N. norvegicus* are usually found in cohesive mud, where they construct and occupy tunnel systems from which they periodically emerge nocturnally to forage (although females rarely emerge when carrying eggs). This is unlike the fully epifaunistic *P. borealis* and *P. camtschaticus*, which congregate in large numbers as protection from predator attacks (Shumway et al. 1985; Hunter 1999; Stevens 2014).

Since the 1970s, there has been much speculation about the location of the settling and nursery habitats of commercially important species because of the risk of overexploitation and the need to develop conservation measures. Several laboratory studies have been conducted over the last 50 years examining the survival and growth of newly settled *H. gammarus* and *H. americanus*, but less is known from field studies, especially for *H. gammarus* but also for the deeper-living juveniles of *P. borealis* and *N. norvegicus* and the non-commercial *G. strigosa*.

The early life stages of *H. americanus* were equally unknown until the 1980s, when the newly settled juveniles were found in mixed and rocky cobble nursery grounds in shallow waters, from the lower intertidal zone to a few meters (rarely up to 80 m) down (Cobb 1971; Richards and Cobb 1986; Barshaw and Bryant-Rich 1988; Cobb and Wahle 1994; Lawton and Lavalli 1995; Cowan 1999). Settled juveniles are hardly ever seen until they are several years old, with a CL of more than 35 mm. In Europe, efforts in the early years to find *H. gammarus* juveniles failed (Howard and Bennett 1979; Howard and Nunny 1981), while later attempts to use

suction sampling and settlement traps to track wild and released juveniles
have not been successful at identifying where this species' juveniles live.
During searches for the smallest *H. gammarus* in European cobble grounds
by similar methods and in habitats preferred by *H. americanus*, none were
found (Linnane et al. 1999, 2001; Mercer et al. 2001; Ringvold et al.
2015). The inability to capture early benthic stage lobsters in cobble was
the case even in the vicinity of sites on known adult lobster grounds where
large numbers of microtagged hatchery-reared lobsters were released,
either directly onto the seabed by divers (Bannister and Addison 1998) or
seeded from the surface in shallow waters (van der Meeren and Næss
1993); however, five to eight years later, these juveniles did apparently
recruit in significant numbers to commercial lobster catchers at those same
sites (Bannister and Addison 1998; Agnalt et al. 1999, 2004). The
inference is that the hatchery-reared juveniles adopted the same cryptic
behaviors as wild lobsters. During 30 years of lobster research, only three
anecdotal observations of possible juvenile lobsters with reasonable
certainty are known by the authors, with no patterns (see section 3.1).

The species richness of potential predators of lobsters (Figure 5) in
Europe is very high (van der Meeren 2000; Robinson and Tully 2000;
Mercer et al. 2001), and includes such highly efficient predatory fishes in
the family Labridae as the ballan wrasse *Labrus bergylta* Ascanius, 1767,
cuckoo wrasse *Labrus mixtus* L., 1758, goldsinny *Ctenolabrus rupestris*
(L., 1758), and several other species of Labridae, as well as gadoid species,
particularly the Atlantic cod *Gadus morhua* L., 1758, several species of
sculpins, (Cottidae), and flatfishes (Pleuronectiformes) (Barshaw and
Bryant-Rich 1988; Wahle et al. 2013b). In addition, octopuses are efficient
predators of decapods, like the globally distributed *Octopus vulgaris*
(Cuvier, 1797), and also cuttlefish (*Sepia* spp.), although this genus does
not occur along the North America Atlantic coast (Young et al. 1998; Jereb
et al. 2015). These visual predator species are present, often in high
densities, over the entire range of *H. gammarus* (Froese and Pauly 2018),
but are not present or are found at lower densities in American waters

(Barshaw and Bryant-Rich 1988; Wahle et al. 2013b). Release trials have shown that all of these fishes would predate on newly released or tethered *H. gammarus* up to a size of 35 mm CL within the first day after release (van der Meeren, 2000; Ball et al. 2001; Mercer et al. 2001). MAFF divers saw that during lobster enhancement releases when the water was warm enough for them to be active, hatchery-reared juveniles could only avoid predators by seeking shelter as quickly as possible (Howard 1983). Released lobsters were not seen in the gut contents of predators after the first 24 h from release, or seen by divers within the first year after release (van der Meeren and Næss 1993). However, recaptures have shown that sufficient numbers of juveniles can survive to boost local stocks if they are released after eight to 10 months in a hatchery and when predation pressure is at the lowest, such as in the early spring (van der Meeren 2001, 2005; Agnalt et al. 2004). Hardly any marine areas in the world have been better investigated than the European coasts, and yet intensive suction sampling and settlement trap deployments in selected, high-density adult lobster grounds with documented recruitment of emergent lobsters have still been unsuccessful in locating small juveniles. However, stage I and II larvae are easily captured, although in low numbers, in light traps and plankton net hauls during a few weeks in the late summer in the coastal waters of Sweden (Öresland and Ulmestrand 2013) and Norway (GI van der Meeren, pers. obs.). Laboratory-reared juveniles show clear selection for specific habitats in release trials, and should thus be even more concentrated after settling than during the planktonic larval stages, yet they have not been found. However, these attempts have revealed that areas with the pebble and cobble substrates in shallow waters that are preferred by the early post-settlement life stages of *H. americanus* are also settling grounds for many decapod species in Europe. This was exemplified by the West Norwegian sampling results from the Lobster Ecology and Recruitment (LEAR) project (Mercer et al. 2001), which is described more below (Figure 6).

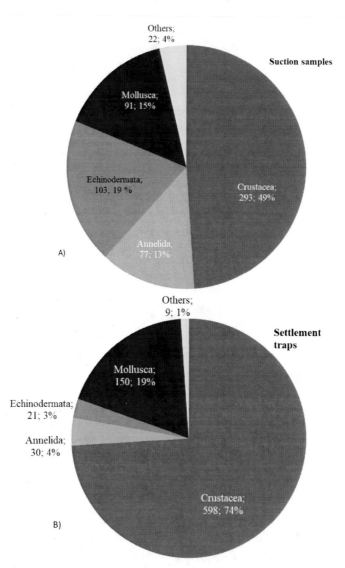

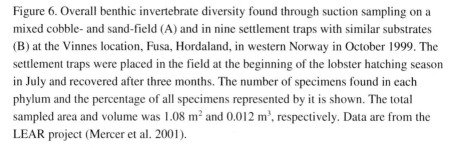

Figure 6. Overall benthic invertebrate diversity found through suction sampling on a mixed cobble- and sand-field (A) and in nine settlement traps with similar substrates (B) at the Vinnes location, Fusa, Hordaland, in western Norway in October 1999. The settlement traps were placed in the field at the beginning of the lobster hatching season in July and recovered after three months. The number of specimens found in each phylum and the percentage of all specimens represented by it is shown. The total sampled area and volume was 1.08 m^2 and 0.012 m^3, respectively. Data are from the LEAR project (Mercer et al. 2001).

The decapods found among cobble during the LEAR project represented not only a high number of species, but also high numbers of individuals, in both suction sampling and settlement traps in natural and lobster fishing areas. Whereas *H. americanus* dominates cobble habitats in North America, a mix of shrimps and several galathean species in all life stages dominated similar habitats in European waters (Figure 7). Small settlement traps provided with cobble and mixed sand and cobble substrates attracted mostly shrimps and juvenile galatheans and brachyurans. Almost 200 small-sized shrimps and nearly 100 galathean juveniles were found per m². Although too small to be predators of settling lobsters, both of these group could compete with them for space and food. There are also considerable gaps in our knowledge on the early life stages of galatheans and non-commercial brachyurans. Therefore, unpublished data and results from Norwegian experiments conducted as parts of the LEAR project were used in the following sections to illuminate some of the similarities and differences in habitat and behavioral traits between laboratory- or hatchery-reared juvenile *H. gammarus* and selected species commonly found in European cobble grounds at 2-10 m depth.

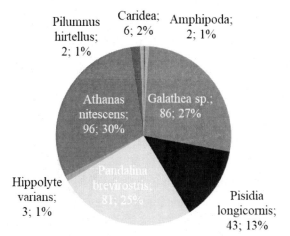

Figure 7. Crustacean diversity in settlement traps recovered in Norway in October 1999. The number of species found in each taxon and the percentage of all specimens represented by it is presented. The total sampled area and volume was 1.08 m² and 0.012 m³, respectively. Data are from the LEAR project (Mercer et al. 2001).

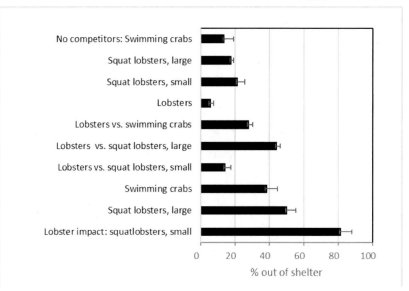

Figure 8. Percentage of time observed out of shelter of newly settled *Homarus gammarus* (L., 1758) with or without different competitors, including *Liocarcinus navigator* (Herbst, 1794) [as *Liocarcinus arcuatus*] (swimming crabs) and/or *Galathea squamifera* Leach, 1814 (squat lobsters), that were either slightly smaller (small) or equal in size to or slightly larger (large) than the lobsters. All specimens used were in the same size range (total length/width: 35 to 50 mm) and were held under a 12/12 h day/night cycle while being observed and counted every 3 h over a 72-h period, in three replicate tanks. The number of lobsters used per tank was 15 individuals, with 15 individuals of each competitor species added per tank as treatments. Data are from the LEAR project (Mercer et al. 2001; Koponen 2003).

A. nitescens, the most common shrimp in the settlement traps and the species most closely resembling stage V of *H. gammarus*, was used to investigate whether and how potential competitors impacted stage V juveniles of *H. gammarus*. One hundred stage V lobsters were paired with 75 *A. nitescen*, sharing space and food in 100 meshed cages (length 135 mm, diameter 70 mm) left on the sea bottom for four months. The 75 lobsters sharing their cages had similar survival (32%) and growth (23.5% in controls and 24.6% overall) to control lobsters, while the weight gain was lower in the shared cages than in the controls (42.2 g in controls vs. 31.3 g overall) (data from the LEAR project: Mercer et al. 2001).

Figure 9. Typical use of biotope (from upper-left to lower-right): *Homarus gammarus* (L., 1758) alert when exposed (left) and in shelter (right), *Nephrops norvegicus* (L., 1758), *Galathea strigosa* (L., 1761), male *Cancer pagurus* (L., 1758) protecting mate, *Carcinus maenas* (L., 1758) partially burrowed in sand. Photographer: AK Woll.

When facing commonly found decapod species in shallow-water cobble grounds in Europe with relatively similar diurnal cycles and shelter preferences to them in the laboratory, naïve newly settled *H. gammarus* were quite effectively able to win and protect covered burrows from potential competitors (Figure 8). The burrowing behavior of *H. gammarus* juveniles was previously described by Berrill (1974). Although lobsters were significantly less affected by their competitors than their competitors were by them (Mann-Whitney U-test: U = 16.5, n_1 and n_2 = 12, p < 0.002), lobsters were out of their shelter significantly more often during the daytime with competitors (e.g., when faced with small squat lobsters: 12.8 ± 3.14% more European lobsters were out of shelter vs. controls; χ^2 = 5.99, df = 1, p = 0.014). Galatheans and brachyurans can find other shelter options that are unavailable to lobsters. The relatively wide- and flat-bodied crabs can dig into the substrate, while the galathean species can use their hooked claws to cling upside-down to the undersides of rocks and other hard objects (Figure 9). However, the presence of slightly larger competitors still led to more lobsters being out of their shelters, which in the field would expose them more to fish predators (Mercer et al. 2001; Koponen 2003) (Figure 8).

Both homarid species perform very similarly in laboratory settings, but to survive settlement and shelter competition in European waters, *H. gammarus* juveniles in the field may have developed antipredator and shelter selection choices that differ from those of *H. americanus* (Wahle and Steneck 1992; Lawton and Lavalli 1995). The competition for shelter and food in Norway, as in other European cobble-bottomed areas, must be very different from that in American waters (Mercer et al. 2001; Wahle et al. 2013a; Ringvold et al. 2015).

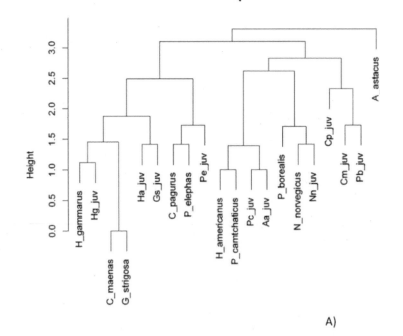

dat[, c((10:12), (21:22))]
Agglomerative Coefficient = 0.59

Figure 10. (Continued).

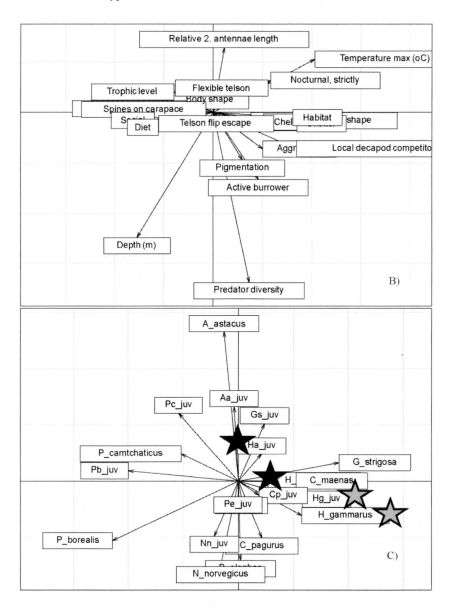

Figure 10. Analyses of similarities and differences among the preferred habitats of all the tested decapods through clustering by the 'Agnes' method (A, previous page) and principal component analyses (PCAs) of the descriptive variables (B) and species (C) along the principal component axes 3 (horizontal) and 5 (vertical). *Homarus gammarus* (L., 1758) are marked with gray stars and *H. americanus* H. Milne Edwards, 1837 are marked with black stars. Statistics were performed by Francois Besnier, IMR.

Statistical comparisons of the known and anticipated traits connected to the habitat of homarid species were made using the 'Agnes' clustering analysis and PCA on data for their habitat traits (Figure 10A). *H. gammarus* were placed in a separate branch from adult *H. americanus*, with juvenile *H. gammarus* clustered even more distantly from the juvenile *H. americanus* (Figure 10A). The PCA of descriptive functional traits showed that these species mainly differed along the PC axis that corresponded with the 'Predator diversity', 'Active burrower', 'Pigmentation', and 'Relative 2 antenna length' traits (Figure 10B). The similarities detected in the juvenile and adult *H. gammarus* may have been an artefact caused by the assumed juvenile biotope and optimal habitat being based on observations of juveniles reared in the laboratory and studied in semi-natural habitats (Figure 10C). Since the antennal length, pigmentation, and burrowing habits are known to be similar between homarid lobster species, at least in laboratory studies, the most significant differences in these analyses were thus caused by differences in predator diversity and decapod competitor diversity (Figure 10B) (van der Meeren 2000; Mercer et al. 2000). *H. gammarus* seems to have biotope traits more in line with those of *C. maenas* and *G. strigosa*, which live in shallow waters with high predation pressure and many decapod competitors, as these species were all clustered quite closely in Figure 10C; the adult *H. americanus* was also clustered within this type of habitat.

It is possible to interpret relative abundances of different species in different habitats and how this may impact juvenile lobster settlement and survival in two ways. From one point of view, one might conclude that European lobster juveniles are 'suppressed' by these 'ecological competitors'. On the other hand, the sparse lobster density and cryptic life cycle may be a low-density strategy that has evolved to reduce the prospects of them having to face predation and competition. The outcome is the same, but without careful experiments it is impossible to determine which one is the driver.

3.5. Behavior

The behaviors (Figure 11) of juvenile and adult *H. americanus* have been described in a number of studies (Cobb 1971; Stewart 1972; Barshaw and Bryant-Rich 1988; see also Atema and Voigt 1995, and references therein).

A behavioral trait that is clearly different among the selected species is seasonal migration. Some female *P. elephas* and *C. pagurus*, but not all, migrate as part of their recruitment strategy (Ansell and Robb 1977; Bennet and Brown 1983; Howard 1982; Follesa et al. 2009; Giacalone et al. 2015), and *P. camtshaticus* and some *H. americanus* migrate to deeper waters during the winter season (Stewart 1972; Sundet 2014). However, *H. gammarus*, *G. strigosa*, *N. norvegicus*, *C. maenas*, and *A. astacus* do not migrate seasonally, and are resident in the same areas throughout the year (Edwards, 1958; Klein Breteler 1976; Chapman 1980; Smith et al. 1998; Ulmestrand and Eggert 2003; Moland et al. 2011; Füreder 2016). Still, *H. gammarus* were found emigrating from artificial shelters over time, and the extents of the home ranges of large lobsters may have been previously underestimated (Jensen et al. 1993; Thorbjørnsen et al. 2018).

The homarid species avoid social interactions other than mating. They use body language and olfactory cues to signal dominance and mating and individual status (Atema and Voigt 1995; Skog 2009). Aggression seems to be negatively correlated to the body/chelae size, where female *H. gammarus* are more aggressive than males, and male *H. gammarus* are more aggressive than male *H. americanus* (van der Meeren et al. 2008; Skog 2009). *H. gammarus* does not seem to respond with decreased aggression upon receiving submissive signals from a competitor: After fights between *H. americanus*, the winner will usually ignore a defeated lobster that tucks in its tail and lowers its chelae, but in *H. gammarus* the winner will continue to chase and attack the opponent as long as they are in the same tank (van der Meeren et al. 2008).

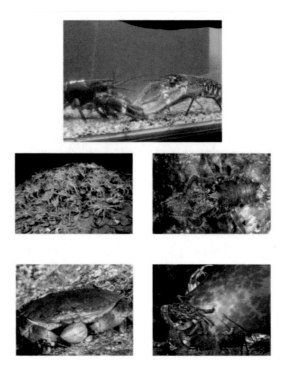

Figure 11. Interactive behaviors in (from top to bottom): male *Homarus gammarus* (L., 1758) (left) and *Homarus americanus* H. Milne Edwards, 1837 (right), fighting; *Paralithodes camtschaticus* (Tilesius, 1815) podding; *Galathea strigosa* (L., 1761) clinging (mating), *Cancer pagurus* L., 1758 (left) and *Carcinus maenas* (L., 1758) (right) foraging. Photographers: homarid lobsters by GI van der Meeren, the rest by AK Woll.

Like *H. americanus*, adult *H. gammarus* are nocturnal predators and scavengers (Smith et al. 1998) with very flexible bodies that can easily move both forwards and backwards. As nocturnal foragers, they shelter in caves and self-made burrows during daylight hours. High-density patches of *N. norvegicus* are usually found in cohesive mud, where they construct and occupy tunnel systems from which they periodically emerge nocturnally to forage, although females rarely emerge when carrying eggs. Naïve, hatchery-reared, one-year-old *H. gammarus* show responses to new environmental factors that are similar to those of wild, adult lobsters, which are active burrowers, nocturnal, shelter-seeking, solitary, and aggressive (van der Meeren 1993) (Figure 12). Whether this also applies to wild juveniles is not known. Homarid lobsters, even when reared in hatcheries, are observed to burrow when they have the opportunity, and are well-suited to life in modified interstitial spaces within rocky and complex bottoms, from which they remove sand and build up with rocks to make entrances that are easy to protect. Their antennae provide them with excellent olfactory and tactile senses for navigating their dark world (Cobb 1971; Atema and Voigt 1995).

Although *C. pagurus* and *C. maenas* are sometime seen buried in sand, only the three nephropid lobsters actually dig and shape shelters, such as by making tunnels underneath rocks or in compact sediments. These are all examples of these species' functional morphology and habitat ecology. The crabs are protected by their bulky carapace and may have less need for full body-covering shelters than the less armored homarid species. In laboratory experiments in which sand-covered tank bottoms are offered and used by crabs, *H. gammarus* juveniles usually prefer to dig underneath artificial shelters put in the tank, while galatheans will use natural crevices in the provided shelters (pers. obs. by the authors).

P. borealis, *P. camtschaticus*, *C. maneas*, and *C. pagurus* do not seek body-covering shelters, but also show other differences from shelter-making lobsters, including camouflage pigmentation and protective behavioral traits. *P. borealis* are good swimmers that live in dense shoals (Shumway et al. 1985). Crabs are often found on softer bottoms, near or in crevices in kelp and alga-covered substrates, environments where they can cover themselves with sand, cling to the surface, or wedge themselves inside crevices. Their pereiopods can fold underneath the body and provide a strong grip on the substrate. *P. camtschaticus* and *P. elephas* are both afforded protection from predators by forming social aggregations and possessing spiny carapaces and long appendages. *G. strigosa* are less social, but still not aggressive like *H. gammarus*. Due to their both flattened and elongated body shape, with a flexible tail tucked underneath the body, and the hooked claws present on all their pereiopods their grip on rocky surfaces and other objects is strong (Figure 9).

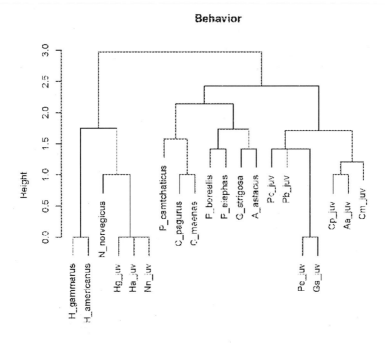

Figure 12. Analyses of similarities and differences between all the selected behavioral traits of the tested decapods ('Agnes' method). Statistics were performed by Francois Besnier, IMR.

Decapod species are found at different depths, from the upper intertidal zone (*C. maenas*), to the upper subtidal zone (*G. strigosa* and *A. astacus*), relatively shallow waters (*H. gammarus*, *C. pagurus*, and *P. elephas*), and deeper waters (*P. borealis* and *N. norvegicus*); some also occupy a wider depth range, from shallow waters down to more than 100 m depth (*P. camtschaticus* and *H. americanus*). Living in the littoral zone requires appropriate responses to the tide. The behavior of *C. maenas* is connected to the tidal cycle, and crabs will move into a shelter when the tide is going out (Crothers 1967, 1968). Typically, they hide under seaweed or rocks, staying humid and sheltered from direct sunlight. In the competition experiments done during the LEAR project, each test was finished by emptying the 0.9 m deep tanks with the experimental animals still in place.

In this process, it was observed that the *H. gammarus* juveniles were all hiding underneath the cobble shelters, while the other species sampled from natural cobble grounds at 4 to 10 m depth all reacted by fleeing their shelters, moving out into full exposure on the open sand. The lobsters behaved exactly as the intertidal *C. maenas* would be expected to do (pers. obs. by the authors).

For the early life stages of *H. gammarus*, no previous attempts to document their natural living biotope has yet succeeded, and thus there are no published observations of the behavior of wild juveniles. After more than a century of studies of hatchery-reared lobsters and juvenile releases to the sea, our present knowledge of the juveniles of this species all comes from laboratory studies. Anecdotal information and observations of laboratory-hatched young-of-the-year juveniles in tanks exist, as described earlier in this chapter (section 3.1), which provide varying levels of detail for habitat reconstruction. Searches in nature for this life stage have been futile thus far. Even after releases of thousands of juveniles from one to 12 months old, very few are seen and none are recaptured in the next three to five years, after which they reappear in traps as adults, in the same location where released (Bannister and Addison 1998; van der Meeren 2003; Agnalt et al. 2004). Notably, a Swedish study suggested that lobster larvae may be retained in the area where they are hatched, which implies that they do seek shelters near to the adults' habitats (Öresland and Ulmestrand 2013). Releases and searches for wild juveniles have been based on knowledge of the habitat of larger lobsters more than 50 mm CL in size and biotopes known to attract *H. americanus* juveniles. However, the lack of actual knowledge of wild *H. gammarus* juveniles' habitat preferences and behavior in the rich biodiversity of European waters make proper monitoring of stock recruitment and conservation measures, including recruitment protection, difficult.

4. DISCUSSION: THE FULL PICTURE

There has been substantial speculation regarding how the juveniles of *H. gammarus* manage to remain elusive. Are the juveniles suppressed by competitors and predators and therefore sparsely distributed, or is the low density of *H. gammarus* versus *H. americanus* a strategy to reduce the prospects of competition and predation? Three main hypothesis have been suggested, that they: (1) stay in deeper waters, (2) settle too sparsely to be detected, or (3) dig deep or hide deep in the substrate, in branched and long burrows or caverns (Mercer et al. 2001).

A bigger-picture perspective is needed to pinpoint how the young of *H. gammarus* differ from those of their closest relative. To evaluate the three hypotheses suggested above, a full ecological picture of this species and others, including their morphology, life history, habitat, and behavior, may be useful. In this chapter, the selection of Northern Atlantic decapod species examined all had several features in common with *H. gammarus*: they need protection when molting, are mainly nocturnal, more or less omnivorous, and live within the same temperature range. However, the species differ from each other in their preferences for habitat depth, bottom quality, and the temperature range they prefer. They also differ in how they avoid predators, the extent of parental care, the duration of the larval period, and in morphology. The most temperate species, *P. elephas*, and the polar *P. camtschaticus* represented the extremes of both temperature preferences and morphology considered, although both share part of their temperature tolerance zone with the homarid lobsters. *P. camtschaticus*, *N. norvegicus*, and *P. borealis* are all commonly found below 100 m depth, well below *H. gammarus*, which is usually found above 70 m depth. These deep-living species have typically reddish pigmentation, less developed chelae, and weaker pereiopods that reflect their life in waters deeper than the sunlit intertidal zone, with reduced diversity of predators and competitors, and shelter from predators in self-made burrows in cohesive mud (*N. norvegicus*) or by gathering in large numbers (*P. camtschaticus* pods, *P. borealis* shoals). The pigmentation, behavior, and habitat of adult *H. gammarus* overlaps with the shallow-living marine species and even the limnic *A. astacus*. Adult lobsters stay in shallow waters even more in *H. americanus*, which are even more tolerant to low temperatures than *H. gammarus* in all life stages; they become inactive in waters colder than 7°C, and temperatures below 5°C inhibit larval molts (Templeman 1936; Nicosia and Lavalli 1999; Wickings and Lee 2002). Still, hybridization of *H. americanus* females with *H. gammarus* males has been documented in European waters (Jørstad et al. 2011; Öresland et al. 2017). Without doubt, *H. gammarus* belongs to the group of shallow-dwelling species.

The natural living biotope of all animals is functionally reflected in their morphology, life history strategy, habitat selection, and behavior. In *H. americanus*, density-dependent population regulation is likely (Fogarty and Idoine 1986; Ennis and Fogarty 1997; Wahle et al. 2013b), while this is not expected in *H. gammarus* since it is rarely found in densities as high as *H. americanus* (Wahle et al. 2013a). Still, habitat has impacts on the size composition of *H. gammarus* (Howard 1980), as well as conspecific chemical stimuli and interspecific interference competition (Wickins and Lee 2002). Wild-caught post-larval *C. maenas* respond to density-dependent cues that lead to slightly faster growth and the development of stronger chelae, increasing their competitive ability (Duarte et al. 2014).

As nocturnal foragers, adult *H. gammarus* are photophobic and darkly pigmented, fast-moving and highly mobile, and shelter during the daytime in caves and burrows that they usually modify and manipulate to provide better protection. In the northern part of their distribution, *H. gammarus* are black to nearly black in color. Their very dark colors are a good fit for living on rocky and complex bottoms, while their long antennae provide them with excellent senses of smell/taste and touch for navigating in their dark world, including when sitting inside deep caverns and tunnels. In many ways, *H. americanus* is similar in morphology to *H. gammarus*, although with slightly lighter pigmentation, which may indicate that their ecological niches are comparable but not exactly the same. In contrast, *P. elephas* adopts communal living in groups for protection, and is not found inside deep caves (Eggleston and Lipcius 1992; Diaz et al. 2001; Goñi and Latrouite 2005; Buscaino et al. 2011). The crayfish *A. astacus* is as dark as *H. gammarus*, but lives in lakes with no crustacean competitors and only fish and bird predators, where it shelters in natural crevices and protects the brood until the juveniles have completed the second post-hatch molt, making them large enough to be able to flee from visual fish predators (Stein and Magnusson 1976; Söderbäck 1994; Füreder 2016).

The ways in which morphology constrains how decapods can move are also telling. *P camtschaticus* move freely in open habitats, but cannot lift their own weight out of water, tail-flip away from danger, or fold up to effectively cling to the substrate. However, they do not need to use these techniques because they instead obtain protection by moving in pods and outgrowing their predators (Dew 1990). Young ones, which are not protected by size, stay in social groups and are covered in pointed spines all over the carapace. *Pandalus borealis*, which is small in size and unable to grip anything, relies on tail-flip escape behaviors, camouflage pigmentation, and hiding in dense shoals above the seabed for protection. In shallower waters, wide-bodied crabs have relatively hard carapaces, no tail joints and strong chelae and pereiopods, making them increasingly difficult to break and swallow for predators as they grow. Young stages of crabs rely on digging shallow depressions in sandy bottoms, hiding in inaccessible crevices, and a strong grip achieved by folding the chelae and pereiopods around kelp stems or other objects to which they can cling, leading to them having a preference for sheltering in kelp forests or other places with algal cover or complex surfaces (Howard 1982; Fernandez et al. 1993). Both *G. strigosa* and *P. elephas* exhibit typically elongated 'lobster' shapes, with quite strong walking pereiopods, and use tail-flips for escape. (Robinson and Tully 2000; Linnane et al. 2001; Ringvold et al. 2015). However, galatheans are more dorsoventrally compressed than the cylindrical homarid lobsters and *A. astacus*. Like brachyurans, galatheans can fold their pereiopods underneath the body and get a strong grip on the surface, as well as cling upside-down by the hooked claws on their strong legs and pointed claws. Lobsters cannot fold their legs underneath the body, but can fold them tightly up on each side of the body. They cannot cling to or grip the substrate or objects, but are shaped well for moving through narrow passages and pushing out sediments from burrows or tunnels. In emerging and adult homarid lobsters, this is the preferred habitat in laboratory studies (Linnane et al. 1999; Jørstad et al. 2001; Wahle et al. 2013a). Even if no *H. gammarus* juveniles have been found in cobble grounds, and extremely few have been found at all, the latter are found in habitats with empty shells or rocky reefs mixed with sandy spots, which can usually be improved by manipulating the entrance and shelter shape. Hatchery-reared *H. gammarus* juveniles released in laboratory tests will shelter faster when the water carries predator odors (Longva Nilsen 2007; Aspaas 2012) and show reduced aggression when a fish is present (van der Meeren 1993), showing how innate responses are still present in intensively cultivated juveniles.

Interestingly, a reaction to water drainage commonly found in intertidal species seems to also be a natural reaction in laboratory-reared *H. gammarus* juveniles. Water drainage induces them to move into drained, but humid and shaded, shelters, whereas galatheans, one of the most common benthic decapod groups in European shallow waters, flee their shelters in the absence of water. Thus, *H. gammarus* acts in the same way as the intertidal *C. maenas*. This could indicate that *H. gammarus* may have evolved to live in the upper littoral zone, perhaps even in the intertidal zone. The nursery grounds of *H. americanus* are shallow as well, but not usually intertidal (Cowan 1999, Wahle et al. 2013b). The innate reaction of *H. gammarus* to shelter in dark and moist places when left out of water indicates that the hypothesis that this species' juveniles have been hard to find because they avoid competition by growing up in deeper waters is not likely to be correct.

Looking at all traits together, the most similar species of those examined are the homarid lobsters (Figure 2 and Tables 1-5), even if they differ in terms of some characteristics of their microhabitats and ecology. In terms of habitat, the rich biodiversity of Europe most probably led to the differences found in juvenile, and to some extent adult, *H. gammarus* and *H. americanus* behavioral, morphological, and life history traits. In Late Medieval Scandinavian literature, before *H. gammarus* was fished commercially, Magnus (1556) and Pedersøn Friis (1599) claimed that foxes, ravens, otters, and men caught small lobsters at low tide by turning stones in the intertidal zone (as cited in Spanier et al. (2015)). No intertidal lobsters have been reported in later centuries. While the fiercest competitor any young lobster will meet is another lobster, the density and biodiversity of other species of competitors and predators is higher in European waters than that found along the North American east coast (and in the biotopes of *H. americanus*). These differences between regions could be the cause for some of the noticeable dissimilarities between *H. gammarus* and *H. americanus*. *H. gammarus* lobsters are more aggressive, showing no tolerance to other lobsters except during mating (van der Meeren et al. 2008; Skog 2009). In *H. gammarus*, each egg is larger and provided with more yolk, causing fewer eggs to be produced per clutch, while the newly hatched larvae are larger and start metamorphosis at larval stage III, leading to them becoming more developmentally advanced stage IV and post-larvae than those of *H. americanus* (Aiken and Waddy 1989; Agnalt 2008; Rötzer and Haug 2015). Larger larvae and juveniles are better prepared to fight for shelter and survival after settlement (Wahle and Steneck 1992).

Although *H. gammarus* is known to live in less dense populations, the catch-per-unit-trap in some locations of *H. gammarus* can be at the same level as that of *H. americanus* (Jørstad et al., unpublished data). The *H. gammarus* stock in the UK and Ireland has been stable and has supported a commercially healthy fishery since 1910 (CEFAS 2014). Stage I and II larvae are easily caught in plankton nets or light traps if these are deployed at hatching time and the following few weeks (Öresland and Ulmestrand 2013). Considering all of the fisheries and field research undertaken along the European coasts, as well as failures to recapture any of the newly released hatchery-reared lobsters that turn up years later in commercial catches (Bannister and Addison 1998; Agnalt et al. 2004), it is not likely that scarcity is the reason that juvenile European lobsters have not yet been found. The UK *H. gammarus* stock is stronger than ever, sustaining a healthy fishery, indicating that nursery ground availability is not limiting the stable recruitment of new lobsters to this stock, which probably has also benefitted from warmer summer ocean temperatures in recent years (CEFAS 2014; Selim et al. 2016). In Scandinavia, it is clear that although the warming seas should be beneficial for the lobsters in this northern portion of their distribution, overexploitation combined with low summer sea temperatures in the 1960s and 1970s resulted in stock collapse (Pettersen et al. 2009). The main cause for the still very low stock in areas outside of lobster marine protected areas, despite benign climate conditions, is still overexploitation (Kleiven et al. 2012; Moland et al. 2013; Thorbjørnsen et al. 2018). The overall picture discussed here indicates that a lack of nursery grounds for juveniles is not plausible, as their innate traits seem to keep them out of the microhabitats of their competitors and common predators.

The full picture points to *H. gammarus* juveniles, much like the adults, living in shallow waters, where they are well-dispersed but not necessarily rare. The cryptic life habit of *H. gammarus* for the first years of its life may have evolved in shallow, highly competitive, and predator-dense biotopes, as a strategy to reduce pressures from competitors and predators by occupying and manipulating their shelters and spaces to keep out predatory or competing species. Thus, the advanced larval metamorphosis, high level of aggression, solitary shelter occupation, and active digging and manipulation of the habitat of *H. gammarus* may all be traits that provide its juveniles access to unique shelters in shallow waters that are so successful they are also unapproachable by humans by all approaches utilized so far.

ACKNOWLEDGMENTS

This work is dedicated to John Mercer, for being an inspiring colleague, mentor, and friend. Thanks are given to Francois Besnier for doing the statistical analyses. The text and content has been greatly improved by additional information from the UK, general and specific inputs and critical comments by Colin Bannister, and suggestions from the editor and a reviewer. All these also assisted in needed language improvements. Without the assistance and collaboration of Hannu Kopponen, Haldis Ringvold, and Vidar Wennevik, we would have had even less information on the juvenile *H. gammarus*. The many and good discussions we had with Rick Wahle, Kari Lavalli, Stan Cobb, and many more lobster friends have been vital for the insights and shared knowledge presented in this chapter. We are also grateful for the exchange of knowledge with our Norwegian and Swedish lobster colleagues, Michaela Aschan (UiT) for shrimp literature, and financial support from the Institute of Marine Research (IMR), Norway. Finally, without good dive buddies, the photos could not have been taken and made available for this chapter. In particular, we thank Nils Aukan, Finn Refsnes, and Øystein Paulsen, who provided additional photos.

REFERENCES

Agnalt A-L (2008) Fecundity of the European lobster (*Homarus gammarus*) off southwestern Norway after stock enhancement: do cultured females produce as many eggs as wild females? *ICES J Mar Sci* 65:164-170.

Agnalt A-L, Jørstad KE, Kristiansen T, Nøstvold E, Farsetveit E, Næss H, Paulsen OI, Svåsand T (2004) Enhancing the European lobster (*Homarus gammarus*) stock at Kvitsøy Island: perspectives on rebuilding Norwegian stocks. In: *Stock Enhancement and Sea Ranching: Developments, Pitfalls and Opportunities* (2nd Ed.). Leber KM, Kitada S, Blankenship HL, Svåsand T (eds.). Oxford, UK: Blackwell Science, p. 415-426.

Agnalt A-L, van der Meeren GI, Jørstad KE, Næss H, Farestveit E, Nøstvold E, Svåsand T, Korsøen E, Ydstebø L (1999) Stock enhancement of European lobster (*Homarus gammarus*); a large-scale experiment off south-western Norway (Kvitsøy). In: *Stock Enhancement and Sea Ranching*. Howell B, Moksness E, Svåsand T (eds.). Oxford, UK: Fishing News Books, Blackwell Science, p. 401-419.

Aiken DE, Waddy SL (1989) Maturity and reproduction in the American lobster. *Can Tech Rep Fish Aquat Sci* 932:59-71.

Alexander RMN (1988) The scope and aims of functional and ecological morphology. *Neth J Zool* 31:3–22.

Ansell AD, Robb L (1977) The spiny lobster *Palinurus elephas* in Scottish waters. *Mar Biol* 43:63-70.

Ascanius P (1767) [In: Ascanius P (1767-1805)] *Icones Rerum Naturalium, ou Figures Enluminées d'Histoire Naturelle du Nord. [Images of Nature, or Iluuminated Figures of the Natural History of the North]* Copenhagen, Denmark: Möller. 5 parts, 36 p., 50 pls. [In French].

Aspaas S (2012) *Behaviour of hatchery-produced European lobster (*Homarus gammarus*), comparing conditioned and naïve juveniles.* M.Sc. thesis, University of Bergen, Bergen, Norway, 49 p.

Atema J, Voight R (1995) Behavior and sensory Biology. In: *Biology of the Lobster* Homarus americanus. Factor JR (ed.). New York, NY, USA: Academic Press Inc., p. 313-348.

Bakke S, Larssen WA, Woll AK, Søvik G, Gundersen AC, Hvingel C, Nilssen EM (2018) Size at maturity and molting probability across latitude in female *Cancer pagurus. Fish Res* 205:43-51.

Ball B, Linnane A, Munday B, Browne R, Mercer JP (2001) The effect of cover on *in situ* predation in early benthic phase European lobster *Homarus gammarus. J Mar Biol Assoc UK* 81:639-642.

Bannister RCA, Addison JT (1998) Enhancing lobster stocks: a review of recent European methods, results, and future prospects. *Bull Mar Sci* 62:369-387.

Barshaw DE, Bryant-Rich DR (1988) A long-term study on the behavior and survival of early juvenile American lobsters, *Homarus americanus*, in three naturalistic substrates: eelgrass, mud, and rock. *Fish Bull* 86:789-796.

Bennet D, Brown C (1983) Crab (*Cancer pagurus*) migrations in the English Channel. *J Mar Biol Assoc UK* 63:371-398.

Bergström BI (2000) The biology of *Pandalus. Adv Mar Biol* 38:55-245.

Berrill M (1974) The borrowing behaviour of newly-settled lobsters, *Homarus vulgaris* (Crustacea, Decapoda). *J Mar Biol Assoc UK* 54:797-801.

Bock WJ (1994) Concepts and methods in ecomorphology. *J Bioscience* 19:403-413.

Buscaino G, Filiciotto F, Gristina M, Buffa G, Bellante A, Maccarrone V, Patti B, Mazzola S (2011) Defensive strategies of European spiny lobster *Palinurus elephas* during predator attack. *Mar Ecol Prog Ser* 423:143-154.

CEFAS – Centre for Environment, Fisheries and Aquaculture Science (2014) Lobster (*Homarus gammarus*). *CEFAS Stock Status Report* 2014, 16 p.

Chapman CJ (1980) Ecology of juvenile and adult *Nephrops*. In: *The Biology and Management of Lobsters Volume II: Ecology and Management*. Cobb JS, Phillips BF (eds.). New York, NY, USA: Academic Press, p. 143-178.

Christiansen M (1969) Crustacea, Decapoda Brachyura. In: *Marine Invertebrates of Scandinavia, Vol. 2*. Christiansen, M (ed.). Oslo, Norway: Universitets Forlager, p. 1-143.

Clutton-Brock TH, Godfray JW (1991) Parental investment. In: *Behavioural Ecology: An Evolutionary Approach* (3rd Ed.). Krebs JR, Davies NB (eds.). Oxford, UK: Blackwell Science, p. 7-29.

Clutton-Brock TH, Harvey PH (1984) Comparative approaches to investigate adaptation. In: *Behavioural Ecology: An Evolutionary Approach* (2nd Ed.). Krebs JR, Davies NB (eds.). Sunderland, MA: Sinauer, p. 234-262.

Cobb JS (1971) The shelter-related behavior of the American lobster *Homarus americanus*. *Ecology* 52:108-115.

Cobb JS, Wahle RA (1994) Early life history and recruitment processes of clawed lobsters. *Crustaceana* 67:1-25.

Cooper RA, Uzmann RJ (1980) Ecology of juvenile and adult *Homarus*. In: *The Biology and Management of Lobsters Volume II: Ecology and Management*. Cobb JS, Phillips BF (eds.). New York, NY, USA: Academic Press, p. 97-142.

Cowan DF (1999) Method for assessing relative abundance, size-distribution, and growth of recently settled and early juvenile lobsters

(*Homarus americanus*) in the lower intertidal zone. *J Crustac Biol* 19:738-751.

Crothers JH (1967) The biology of the shore crab *Carcinus maenas* (L.). 1. The background – anatomy, growth and life history. *Field Stud* 2:407-434.

Crothers JH (1968) The biology of the shore crab *Carcinus maenas* (L.). 2. The life of the adult crabs. *Field Stud* 2:579-614.

Cuvier GL (1798) *Tableau Élementaire de l'Histoire Naturelle des Animaux*. [*Elementary Table of the Natural History of Animals*]. Paris, Fance: Baudouin, 710 p. [In French].

D'Abramo LR, Baum NA, Bordner CE, Conklin DE (1983) Carotenoids as a source of pigmentation in juvenile lobsters fed a purified diet. *Can J Fish Aquat Sci* 40:699-704.

D'Agostino A (1980) Growth and color of juvenile lobsters (*Homarus americanus*) kept on diets of natural and artificial foodstuff. In: *Lobster Nutrition Workshop Proceedings*. Bayer RC, D'Agostino A (eds.). *Maine Sea Grant Publ Tech Rep* 58:41-48.

Denney NH, Jennings S, Reynolds JD (2002) Life-history correlates of maximum population growth rates in marine fishes. *Proc R Soc B* 269:2229-2237.

Dew CB (1990) Behavioral ecology of podding red king crab, *Paralithodes camtschatica*. *Can J Fish Aquat Sci* 47:1944-1958.

Diaz D, Mari M, Abello P, Demestre M (2001) Settlement and juvenile habitat of the European spiny lobster *Palinurus elephas* (Crustacea: Decapoda: Palinuridae) in the western Mediterranean Sea. *Sci Mar* 65:347-356.

Drengstig A, Bergheim T, Drengstig T, Kollsgård I,. Svensen R (2003). *Testing of a new feed especially manufactured for European lobster* (Homarus gammarus L.). Report RF-Rogaland Research – 2003/183, 18 p.

Duaerte RC, Ré A, Flores AAV, Queiroga H (2014) Conspecific cues affect stage-specific molting frequency, survival, and claw morphology of early juvenile stages of the shore crab *Carcinus maenas*. *Hydrobiologia* 724:55-66.

Dullemeijr P (1974) *Concepts and Approaches in Animal Morphology.* Assen, The Netherlands: Van Gorcum.

Edwards RL (1958) Movements of individual members in a population of the shore crab, *Carcinus maenas* L., in the littoral zone. *J Anim Ecol* 27:34-45.

Eggleston DB, Lipcius RN (1992) Shelter selection by spiny lobster under variable predation risk, social conditions and shelter size. *Ecology* 73:992-1011.

Ennis GP, Fogarty MJ (1997) Recruitment overfishing reference point for the American lobster, *Homarus americanus. Mar Freshwat Res* 48:1029-1034.

Fabricius JC (1787) *Mantissa Insectorum Sistens Eorum Species Nuper Detectus Adjectis Characteribus Genericis Differentiis Specificis, Emendationibus, Observationibus.* 1. [*Insect Species Newly Discovered by Science with their Characteristics, Generic Differences, Species Emedations, and Observations.* 1]. Hafniae. 348 p. [In Latin].

Factor JR (ed.) (1995) *Biology of the Lobster* Homarus americanus. New York, NY, USA: Academic Press Inc., 528 p.

Fernandez M, Iribarne O, Armstrong D (1993) Habitat selection by young-of-the-year Dungeness crab, *Cancer magister*, and predation risk in intertidal habitats. *Mar Ecol Prog Ser* 92:171-177.

Fogarty MJ (1994) Populations, fisheries, and management. In: *Biology of the Lobster* Homarus americanus. Factor JR (ed.). New York, NY, USA: Academic Press Inc., p. 111-138

Fogarty MJ, Idoine JS (1986) Recruitment dynamics in an American lobster (*Homarus americanus*) population. *Can J Fish Aquat Sci* 43:2368-2376.

Follesa MC, Cuccu D, Cannas R, Cau A (2009) Movement patterns of the spiny lobster *Palinurus elephas* (Fabricius, 1787) from a central western Mediterranean protected area. *Sci Mar* 73:499-506.

Froese R, Pauly D (eds.) (2018) *FishBase.* World Wide Web electronic publication, version (10/2018). Available at: http://www.fishbase.org.

Füreder, L (2016) Crayfish in Europe: biogeography, ecology and conservation. In: *Freshwater Crayfish: A Global Overview.* Kawai T,

Faulkes Z, Scholtz G (eds.). Boca Raton, FL, USA: CRC Press, p. 594–627.

Giacalone VM, Barausse A, Gristina M, Pipitone C, Visconti V, Badalamenti F, D'Anna G (2015) Diel activity and short-distance movement pattern of the European spiny lobster, *Palinurus elephas*, acoustically tracked. *Mar Ecol* 36:389-399.

Goñi R, Latrouite D (2005) Review of the biology, ecology and fisheries of *Palinurus* spp. species of European waters: *Palinurus elephas* (Fabricius, 1787) and *Palinurus mauritanicus* (Gruvell, 1911) *Cah Biol Mar* 46:127-142.

Groeneveld C, Goñi R, Díaz D (2013) *Palinurus* species. In: *Lobsters: Biology, Management, Aquaculture and Fisheries* (2nd Ed.). Phillips BF (ed.). Oxford, UK: John Wiley and Sons Ltd., p. 326-348.

Hansen HØ, Aschan M (2000) Growth, size- and age-at-maturity of shrimp, *Pandalus borealis*, at Svalbard related to environmental parameters. *J Nortw Atl Fish Sci* 27:83-91.

Hanson JM (2009) Predator-prey interactions of American lobster (*Homarus americanus*) in the southern Gulf of St. Lawrence, Canada. *New Zeal J Mar Freshwat Res* 43:69-88.

Harrison MK, Crespi BJ (1999) A phylogenetic test of ecomorphological adaption in *Cancer* crabs. *Evolution* 53:961-965.

Harvey PH, Nee S (1997) The phylogenetic foundations of behavioural ecology. In: *Behavioural Ecology: An Evolutionary Approach* (4th Ed.). Krebs JR, Davies NB (eds.). Oxford, UK: Blackwell Science, p. 334-349.

Herbst J (1794) [In: Herbst J (1782-1804)] *Naturgeschichte der Krabben und Krebse*. [*Natural Histroy of Crabs and Crayfish*]. (3 Vols.) Berlin, Germany: Bei Gottlieb August Lange. [In German].

Herrick FH (1911) Natural history of the American lobster. *Bull US Bur Fish* 29:147-408.

Howard AE (1980) Substrate controls the size composition of lobsters (*Homarus gammarus*) populations. *J Cons int Explor Mer* 39:130-133.

Howard AE (1982) The distribution and behaviour of ovigerous edible crabs (*Cancer pagurus*), and subsequent sampling bias. *J Cons int Explor Mer* 40:259-261.

Howard AE (1983) The behaviour of hatchery-reared juvenile lobster (*Homarus gammarus*), released and observed by divers. *ICES CM* 1983/K:3 (mimeo), 5 p.

Howard AE, Bennett DB (1979) The substrate preference and burrowing behavior of juvenile lobsters (*Homarus gammarus* (L.)). *J Nat Hist* 13:433-438.

Howard AE, Nunny RS (1981) Effects of near-bed current speeds on the distribution and behaviour of the lobster, *Homarus gammarus* (L.). *J Exp Mar Biol Ecol* 71:27-42.

Hunter E (1999) Biology of the European spiny lobster, *Palinurus elephas* (Fabricius, 1787) (Decapoda, Palinuroidea). *Crustaceana* 72:545-565.

Ingle RW (1992) *Larval Stages of Northeastern Atlantic Crabs*. London, UK: Chapman and Hall, 363 p.

Jensen AC, Collins KJ, Free EK, Bannister RCA (1993) Lobster (*Homarus gammarus*) movement on an artificial reef: the potential use of artificial reefs for stock enhancement. *Crustaceana* 67:198-211.

Jereb P, Allcock A, Lefkaditou E, Piatkowski U, Hastie LC, Pierce, GJ (eds.) (2015) Cephalopod biology and fisheries in Europe: II. Species Accounts. *ICES Cooperative Research Report* No. 325, 360 p.

Jørstad KE, Agnalt A-L, Farestevit E (2011) The introduced American lobster, *Homarus americanus* in Scandinavian waters. In: *In the Wrong Place – Alien Marine Crustaceans: Distribution, Biology and Impacts*. Galil BS, Clark PF, Carlton J (eds.). London, UK: Springer, p. 625-638.

Jørstad KE, Agnalt A-L, Kristiansen, TS, Nøstvold E (2001). High survival and growth of European lobster juveniles (*Homarus gammarus*), reared communally with natural bottom substrate. *Mar Freshwat Res* 52:1431-1438.

Kaufman L, Rousseeuw PJ (1990) *Finding Groups in Data: An Introduction to Cluster Analysis*. New York, NY, USA: Wiley.

Klein Breteler WCM (1976) Migration of the shore crab, *Carcinus maenas*, in the Dutch Wadden Sea. *Neth J Sea Res* 10:338-353.

Kleiven, AR, Olsen EM, Vølstad JH (2012) Total catch of a red-listed marine species is an order of magnitude higher than official data. *PLoS ONE* 7:e31216. DOI:10.1371/journal.pone.0031216.

Koponen H (2003) *Interspecific competition among hatchery reared European lobster (*Homarus gammarus *L.) juveniles and wild benthic decapods*. M.Sc. thesis, University of Bergen, Bergen, Norway, 64 p.

Kristiansen T, Drengstid A, Bergheim A, Drengstig T, Kollsgård I, Svendsen R, Nøstvold E, Farestveit E, Aardal L (2004) Development of methods for intensive farming of European lobster in recirculated seawater. Results from experiments conducted at Kvitsøy lobster hatchery from 2000 to 2004. *Fsiken og Havet (IMR)* 6, 55 p.

Krøyer H (1838) Conspectus Crustaceorum Groenlandiæ. [An Examination of the Crustaceans of Greenland]. *Naturhistorisk Tidsskrift* 2:249-261. [In Danish].

Lamarck JBPA de Monet de (1798) [In: Lamarck JBPA de Monet de (1798-1816)]. *Encyclopédie Méthodique. Tableau Encyclopédie et Méthodique de Trois Règnes de la Nature. Vers, Coquilles, Mollusques et Polypiers. [Methodical Encyclopedia. Encyclopedia and Methodical Tables of the Three Kingdoms of Nature. Worms, Shells, Molluscs and Polypiers]*. Paris, France: V. Agasse. [In French].

Lawton P, Lavalli KL (1995) Postlarval, juvenile, adolescent, and adult ecology. In: *Biology of the Lobster* Homarus americanus. Factor JR (ed.). New York, NY, USA: Academic Press Inc., p. 47-88.

Leach WE (1813) [In: Leach WE (1813-1814)]. Crustaceology. *Brewster's Edinburgh Encyclopaedia* 7:383-437.

Lim BK, Sakurai N, Sugihara T, Kittaka J (1997) Survival and growth of the American lobster *Homarus americanus* fed formulated feeds. *Bull Mar Sci* 61:159-163.

Linnaeus C (1758) *Systema Naturae per Regna Tria Naturae, Secundum Classes, Ordines, Genera, Species, cum Characteribus, Differentiis, Synonymis, Locis. [System for Classifying the Three Kingdoms of Nature According to Classes, Orders, Genera, Species, and their*

Characteristics, Differences, Synonyms, and Places]. (10th Ed.). Holmiae : Laurentii Salvii, 824 p. [In Latin].

Linnaeus C (1761) *Fauna Suecica sistens Animalia Sueciae Regni: Mammalia, Aves, Amphibia, Pisces, Insecta, Vermes. Distributa per Classes, Ordines, Genera, Species, cum Differentiis Specierum, Synonymis Auctorum, Nominibus Incolarum, Locis Natalium, Descriptionibus Insectorum.* (Editio altera, auctior). [*The Swedish Fauna, or the Animal Kingdom in Sweden: Mammals, Birds, Amphibians, Fishes, Insects, Worms. Distributed by Classes, Orders, Genera, Species, and the Species' Differences, Preferred Synonyms, Local Names, and Origins, Description of the Insects.* (Revised edition)]. Stockholm: Laurentii Salvii, 578 p. [In Latin].

Linnane A, Ball B, Mercer JP, Browne R, van der Meeren GI, Ringvold H, Bannister RCA, Mazzoni D, Munday B (2001) The search for the early benthic phase (EBP) European lobster: a trans-European study of cobble fauna. *Hydrobiologia* 465:63-72.

Linnane A, Mercer JP, Browne R, van der Meeren G, Bannister C, Mazzoni D, Munday B, Ringvold H (1999) Understanding the factors that influence European lobster recruitment: a trans-European study of cobble fauna. *J Shellfish Res* 18:719-720.

Longva Nilsen K (2007) *Difference in behaviour of the European lobster (*Homarus gammarus*) exposed to homogeneous and heterogeneous environment when supplying ketones from predator.* M.Sc. thesis, University of Bergen, Bergen, Norway.

Maechler M, Rousseeuw P, Struyf A, Hubert M, Hornik K (2018) *Cluster: Cluster Analysis Basics and Extensions.* R Package Version 2.0.7-1. Available at: https://CRAN.R-project.org/package=cluster.

MarLIN – Marine Life Information Network (2016) *Marine Life Information Network.* Plymouth, UK: Marine Biological Association of the United Kingdom. Available at: http://www.marlin.ac.uk.

Mercer J, Bannister RCA, van der Meeren GI, Debuse V, Mazzoni D, Linnane A, Ball B (2000) The influence of competitive interactions on the abundance of early benthic stage European lobster (*Homarus gammarus* L.) and hence on the carrying capacity of lobster habitat.

Final Report LEAR FAIR CT-1775. Carna, County Galway, Ireland: Shellfish Research Laboratory, 158 p.

Mercer J, Bannister RCA, van der Meeren GI, Debuse V, Mazzoni D, Linnane A, Ball B (2001) An overview of the LEAR (Lobster Ecology and Recruitment) project: the results of field and experimental studies on the juvenile ecology of *Homarus gammarus* in cobble. *Mar Freshwat Res* 52:1291-1302.

Milne Edwards H (1837) *Histoire naturelle des Crustacés, comprenant l'anatomie, la Physiologie et la Classification de ces Animaux. Vol. 2. [Natural History of Crustaceans, Comprising the Physiology and the Classification of these Animals. Vol. 2].* Paris, France: Librairie Encyclopédique de Roret. [In French].

Moland E, Olsen EM, Andvord K, Knutsen JA, Stenseth NC, Sainte-Marie B (2011) Home range of European lobster (*Homarus gammarus*) in a marine reserve: implications for future reserve design. *Can J Fish Aquat Sci* 68:1197-1210.

Moland E, Olsen EM, Olsen EM, Knutsen H, Garrigou P, Espegrend SH, Kleiven AR, André C, Knutsen JA (2013). Lobster and cod benefit from small-scale northern marine protected areas: inference from an empirical before-after control-impact study. *Proc R Soc B* 280:2012-2679.

Nelson GJ, Platnick NI (1981) Three-taxon statements: a more precise use of parsimony? *Cladistics* 7:351-366.

Nicosia F, Lavalli K (1999) Homarid lobster hatcheries: their history and role in research, management, and aquaculture. *Mar Fish Rev* 61:1-57.

Nilssen EM, Hopkins CCE (1991). Population parameters and life histories of the deep-water prawn, *Pandalus borealis,* from different regions. *ICES CM Doc*, No K2.

Öresland V, Ulmestrand M (2013) European lobster subpopulations from limited adult movements and larval retention. *ICES J Mar Sci* 70:532-539.

Öresland V, Ulmestrand M, Agnalt A-L, Oxby G (2017) Recorded captures of American lobster (*Homarus americanus*) in Swedish waters and observed predation on the European lobster (*Homarus gammarus*). *Can J Fish Aquat Sci* 74:1503-1506.

Palomares MLD, Pauly D (eds.) (2018) *SeaLifeBase*. World Wide Web electronic publication, version (10/2018). Available at: http://www.sea lifebase.org.

Pearson J (1908) Cancer. *Liverpool Marine Biological Committee Memoirs on Typical British Marine Plants and Animals Volume 16*. Liverpool, UK: University of Liverpool, 263 p.

Poore GCB, Ahyong ST, Taylor J (eds.) (2011) *Crustacean Issues 20: The Biology of Squat Lobsters*. Collingwood, Australia: CRC Press, 363 p.

R Core Team (2016) *R: A Language and Environment for Statistical Computing*. Vienna, Austria: R Foundation for Statistical Computing. https://www.R-project.org/.

Richards RA, Cobb JS (1986) Competition for shelter between lobsters (*Homarus americanus*) and Jonah crabs (*Cancer borealis*): effects of relative size. *J Fish Res Board Can* 43:2250-2255.

Ringvold H, Grytnes J-A, van der Meeren GI (2015) Diver-operated suction sampling in Norwegian cobble grounds: technique and associated fauna. *Crustaceana* 88:184-202

Robinson M, Tully O (2000) Seasonal variation in community structure and recruitment of benthic decapods in a sub-tidal cobble habitat. *Mar Ecol Prog Ser* 206:181-191.

Rötzer MAIN, Haug JT (2015) Larval development of the European lobster and how small heterochronic shifts lead to more pronounced metamorphosis. *Int J Zool* 2015:345172, 17 p. DOI:10.1155/2015/345172.

Sadovy Y (2001) The threat of fishing to highly fecund fishes. *J Fish Biol* 59 (Suppl. A):90-108.

Sastry AN (1983) Ecological aspects of reproduction. In: *The Biology of Crustacea, Volume 8: Environmental Adaptations*. Vernberg FJ, Vernberg WB (eds.). New York, NY, USA: Academic Press Inc., p. 179-269.

Selim SA, Blanchard JL, Bedford J, Webb TJ (2016) Direct and indirect effects of climate and fishing on changes in coastal ecosystem services: a historical perspective from the North Sea. *Reg Env Change* 16:341-351.

Sheehy MRJ, Bannister RCA, Wickins JF, Shelton PMJ (1999) New perspectives on the growth and longevity of the European lobster (*Homarus gammarus*). *Can J Fish Aquat Sci* 56:1904-1915.

Shields JD (1991) The reproductive ecology and fecundity of *Cancer* crabs. In: *Crustacean Issues 7: Crustacean Egg Production*. Wenner AM, Kuris A (eds.). Rotterdam, The Netherlands: A.A. Balkema, p. 193-213.

Shumway SE, Perkins HC, Schick DF, Stickney AP (1985) Synopsis of biological data on the pink shrimp, *Pandalus borealis* Krøyer, 1838. *FAO Fisheries Synopsis* No. 144.

Skog M (2009) Intersexual differences in European lobster (*Homarus gammarus*): recognition mechanisms and agonistic behaviours. *Behaviour* 146:1071-1091.

Smith IP, Collins KJ, Jensen AC (1998) Movement and activity patterns of the European lobster, *Homarus gammarus*, revealed by electromagnetic telemetry. *Mar Biol* 132:611-623.

Söderbäck B (1994) Interactions among juveniles of two freshwater crayfish species and a predatory fish. *Oecologia* 100:229-235.

Spanier E, Lavalli KL, Goldstein JS, Groeneveld JC, Jordaan GL, Jones CM, Phillips BF, Bianchini ML, Kibler RD, Díaz D, Mallol S, Goñi R, van der Meeren GI, Agnalt A-L, Behringer DC, Keegan WF, Jeffs A (2015) A concise review of utilization of lobster by worldwide human populations from prehistory to the early modern era. *ICES J Mar Sci* 72 (Suppl. 1):i7-i21.

Stein RA, Magnusson JJ (1976) Behavioral response of crayfish to a fish predator. *Ecology* 57:751-761.

Stevens BD (ed.) (2014) *King Crabs of the World: Biology and Fisheries Management*. Boca Rota, FL, USA: CRC Press, 608 p.

Stewart LL (1972) *The seasonal movements, population dynamics, and ecology of the lobster,* Homarus americanus, *off Ram Island, Connecticut*. Ph.D. thesis, University of Connecticut, Storrs, Connecticut, USA.

Subramoniam T (2016) Spermatogenesis In: *Sexual Biology and Reproduction in Crustaceans*. Subramoniam T (ed.). New York, NY, USA: Academic Press Inc., p. 293-324.

Sundet J (2014) Red king crab in the Barents Sea. In: *King Crabs of the World: Biology and Fisheries Management*. Stevens BD (ed.). Boca Rota, FL, USA: CRC Press, p. 485-500.

Templeman W (1936) The influence of temperature, salinity, light and food conditions on the survival and growth of the larvae of the lobster (*Homarus americanus*). *J Biol Board Can* 2:485-497.

Thorbjørnsen et al. (2018) Replicated marine protected areas support movement of larger, but not more, European lobsters to neighbouring fished areas. *Mar Ecol Prog Ser* 595:123-133.

Tilesius WL (1815) De cancris Camtschaticis, Oniscus, Entomostracis et cancellis marinis microscopicis noctilucentibus, cum appendice de acaris et ricinis Camtschaticis. [A study of the crabs of Kamchatka, also the isopods, the entomostracans, and the microscopic phosphorescent marine Cancelli with an Appendix concerning the mites and ticks of Kamchatka]. *Mémoires de l'Académie Impériale des Sciences, St. Pétersbourg* 5:331-405. [In Latin].

Tlusty M, Hyland C (2005) Astaxanthin deposition in the cuticle of juvenile American lobster (*Homarus americanus*): implications for phenotypic and genotypic coloration. *Mar Biol* 147:113-119.

Ulmestrand M, Eggert H (2001) Growth of Norway lobster, *Nephrops norvegicus* (Linnaeus 1758), in the Skagerrak, estimated from tagging experiments and length frequency data. *ICES J Mar Sci* 58:1326-1334.

van der Meeren GI (1993) Initial response to physical and biological conditions in naive juvenile lobsters *Homarus gammarus* L. *Mar Behav Physiol* 24:79-92.

van der Meeren GI (2000) Predation on hatchery-reared lobsters *Homarus gammarus* released in the wild. *Can J Fish Aquat Sci* 57:1794-1803.

van der Meeren GI (2001) Effects of experience with shelter in cultivated juvenile European lobsters *Homarus gammarus*. *Mar Freshwat Res* 52:1487-1494.

van der Meeren GI (2005) Potential of ecological studies to improve survival of cultivated and released European lobsters *Homarus gammarus*. *New Zeal J Mar Freshwat Re*s 39:399-424.

van der Meeren GI, Chandrapavan A, Breithaupt T (2008) Sexual and aggressive interactions in a mixed group of lobsters *Homarus gammarus* and *H. americanus*. *Aquat Biol* 2:191-200.

van der Meeren, GI, Næss H (1993) *Lobster* (Homarus gammarus) *catches in Southwestern Norway, including the first recaptures of previously released juveniles*. ICES CM Doc 1993/K:29, 11 p. Available at: http://hdl.handle.net/11250/105188.

Vogt G (2012) Ageing and longevity in the Decapoda (Crustacea): a review. *J Comp Zool* 251:1-25.

Wahle RA, Bergeron C, Tremblay J, Wilson C, Burdett-Coutts V, Comeau M, Rochette R, Lawton P, Glenn R, Gibson M (2013b) The geography and bathymetry of American lobster benthic recruitment as measured by diver-based suction sampling and passive collectors. *Mar Biol Res* 9:42-58.

Wahle RA, Castro KM, Tully O, Cobb JS (2013a) *Homarus*. In: *Lobsters: Biology, Management, Aquaculture and Fisheries* (2nd Ed.). Phillips BF (ed.). Oxford, UK: John Wiley and Sons Ltd., p. 221-258.

Wahle RA, Steneck RS (1992) Habitat restrictions in early benthic life: experiments on habitat selection and *in situ* predation with the American lobster. *J Exp Mar Biol Ecol* 157:91-114.

Wickins, JF, Lee D O (2002) *Crustacean Farming: Ranching and Culture* (2nd Ed.). Oxford, UK: Blackwell Science, 446 p.

Williams DM, Ebach MC (2004) The reform of palaeontology and the rise of biogeography: 25 years after 'Ontogeny, Phylogeny, Paleontology and the Biogenetic Law' (Nelson, 1978). *J Biogeogr* 31:1-27.

Williamson DI (1983) Crustacea Decapoda: Larvae VIII Nephropidea, Palinuridea and Eryonidea. *Fiches D'identification de Zooplancton*, 167/168, 8 p.

Williamson, HC (1905) A contribution to the life history of the lobster (*Homarus vulgaris*). *Ann Rep Fish Board Scotland* 23:65-155.

Young RE, Vecchione M, Donovan D (1998). The evolution of coleoid cephalopods and their present biodiversity and ecology. *S Afr J Mar Sci* 20:393-420.

BIOGRAPHICAL SKETCHES

Gro I. van der Meeren

Affiliation: Institute of Marine Research (IMR), Norway

Education: Dr. Phil. (equal to Ph.D.) at University of Bergen, 2003

Research and Professional Experience: Studied crab and lobster biology, ecology, and behavior since 1984 until 2007. Since 2007 has been working to understand ecological processes, developing assessment tools and status reports for integrated large marine ecosystems, and communicating marine science for governmental, public, and education use. Scuba diver since 1978.

Professional Appointments: Senior researcher (Professor level)

Honors: Best public IMR communicator 2015

Publications from the Last 3 Years:

Bundy A, Chuenpagdee R, Boldt JL, de Fatima Borges M, Camara ML, Coll M, Diallo I, Fox C, Fulton EA, Gazihan A, Jarre A, Jouffre D, Kleisner KM, Knight B, Link J, Matiku PP, Masski H, Moutopoulos DK, Piroddi C, Raid T, Sobrino I, Tam J, Thiao D, Torres MA, Tsagarakis K, van der Meeren GI, Shin Y-J (2017) Strong fisheries management and governance positively impact ecosystem status. *Fish Fisher* 18:412-439.

Coll M, Shannon LJ, Kleisner KM, Juan-Jordá MJ, Bundy A, Akoglu AG, Banaru D, Boldt JL, Borges MF, Cook A, Diallo I, Fu C, Fox C, Gascuel D, Gurney LJ, Hattab T, Heymans JJ, Jouffre D, Knight BR, Kucukavsar S, Large SI, Lynam C, Machias A, Marshall KN, Masski H, Ojaveer H, Piroddi C, Tam J, Thiao D, Thiaw M, Torres MA, Travers-Trolet M, Tsagarakis K, Tuck I, van der Meeren GI, Yemane D, Zador SG, Shin YJ (2016) Ecological indicators to capture the effects of fishing on biodiversity and conservation status of marine exploited ecosystems. *Ecol Indic* 60:947-962.

Garcia-Soto C, van der Meeren GI, Busch JA, Delany J, Domegan C, Dubsky K, Fauville G, Gorsk, G, von Juterzenka K, Malfatti F, Mannaerts G, McHugh P, Monestiez P, Seys J, Węsławski JM, Zielinski O (2017) Advancing citizen science for coastal and ocean research. In: *Position Paper 23 of the European Marine Board.* V Kellett P, Delany J, McDonough N (eds.). Ostend, Belgium: European Science Foundation, 112 p. [In French].

Goodwin LJ, van der Meeren GI (2018) To secure rich and clean oceans people are our greatest resource. *Tvergastein* 11:68-79.

Hansen C, Skern-Mauritzen M, van der Meeren GI, Jänkel A, Drinkwater K (2016) Set-up of the Nordic and Barents Seas (NoBa) Atlantis model. *Fisken og Havet* 2 (2016), 112 p.

Prozorkevich D, Johansen GO, van der Meeren GI (2018) Survey report from the joint Norwegian/Russian ecosystem survey in the Barents Sea and adjacent waters, August-October 2017. *IMR/PINRO Joint Report Series*, No. 2/2018: 100 p.

van der Meeren, GI, Goodwin, LJ (eds.) (2016) Ocean sustainability under global change – top priorities for Norwegian research and prospects for global collaboration. *Rapport fra Havforskningen*, 36-2016, 34 pp.

Astrid K. Woll

Affiliation: Woll Naturfoto; previously Møre Research, retired (2015)

Education: Cand. real. Master of Science in Marine Biology, University in Trondheim. Senior scientist, evaluated by external committee (Ph.D.-level)

Research and Professional Experience: 20 years of research experience at Møre. Research mainly carried out on shellfish. Underwater photographer with approximately 2000 logged dives. Class I working licence (not off-shore) from 1994.

Publications from the last 3 Years:

Bakke S, Larssen WE, Woll AK, Søvik G, Gundersen AC, Hvingel C, Nilssen EM (2018) Size at maturity and molting probability across latitude in female *Cancer pagurus*. *Fish Res* 205:43-51.

Larssen WE, Aas GH, Woll AK (2015) Sammenligning mellom industrielt agn og tradisjonelt seiagn for taskekrabbe. [Comparisons between industrial trap bait and traditional baite (saithe) for Edible crab pots]. In: *Blue Bio-resources*. Gundersen AC, Velle LG (eds.). Stamsund, Norway: Orkana Forlag, p. 135-148. [In Norwegian].

Søvik G, Furevik D, Jørgensen T, Bakke S, Larssen WE, Thangstad TH, Woll AK (2016) The Norwegian *Nephrops* fishery - historic trends and present exploitation and management of a valuable resource. In: *Sustainable Bio-resources: Management, Product Development and Raw Material Quality*. Thu BJ, Gundersen AC (eds.). Stamsund, Norway: Orkana Forlag, p. 95-120.

In: Lobsters

Editor: Brady K. Quinn

ISBN: 978-1-53615-711-6

© 2019 Nova Science Publishers, Inc.

Chapter 3

BIOLOGIC AND SOCIOECONOMIC HARVESTING STRATEGIES FOR THE CARIBBEAN SPINY LOBSTER FISHERIES

Ernesto A. Chávez[*]

Centro Interdisciplinario de Ciencias Marinas,
Instituto Politécnico Nacional, La Paz, B.C.S., México

ABSTRACT

The spiny lobster fisheries exploited by 25 Caribbean countries are subject to heterogeneous harvesting practices, and some of them undergo recurrent socioeconomic crises. Therefore, in this chapter a meta-analysis was conducted with the purpose of providing general management recommendations by evaluating the performance of the five main Caribbean lobster fisheries (those in the Bahamas, Brazil, Cuba, Nicaragua, and the United States of America), as well as the total production of the region. The associated costs, benefits, and social values of a small fishery in southeastern Mexico, in the northwestern Caribbean,

---

[*] Corresponding Author's E-mail: echavez@ipn.mx.

were used as reference values. The stocks were assessed by reconstructing the age structure of each population over a 15-year period, and then the catch, profit, direct provisioning of jobs, and profits per fisher were estimated for each fishery in simulations with different ages of first catch (t_c). Based on these, the fisheries mortality (F) values needed to attain the maximum sustainable yield (F_{MSY}) and the maximum economic yield (F_{MEY}) were selected as the optimum harvesting options for different fisheries. The results showed that the yield increased with the t_c, and in three cases the yields at the F_{MSY} were higher than those at the F_{MEY}. The profits were higher at higher t_c values in three fisheries, meaning that they were more profitable if harvested at their F_{MEY} levels. One fishery (that in the Bahamas) was not profitable if harvested at the F_{MSY} level at any age. The social value of most of the fisheries, calculated as the profits/fisher, was the highest at a t_c of 3 years, and again was higher if the F_{MEY} strategy was applied. However, in the Bahamas' fishery the social value at the F_{MSY} was negative at any t_c. Based on the widespread distribution of this species of spiny lobster across the coasts of the Caribbean and the heterogeneous exploitation practices applied to its fisheries, the creation of a multinational organization in charge of regulating and managing the fishery in each country is recommended to achieve the sustainable exploitation of this resource within the framework of conservation.

Keywords: Caribbean, optimum yield, *Panulirus argus*, spiny lobster

1. INTRODUCTION

Stock assessment has been a major focus of fisheries scientists' efforts for a long time, and the related exploitation of fished stocks to the level of their maximum sustainable yield has been one of the main goals of fisheries science and management (Beverton and Holt 1957; Cushing 1968; Gulland 1969, 1988; Ricker 1975; Hilborn and Walters 1992; Quinn II and Deriso 1999). This approach has the purpose of acting as a "holding action against the forces of resource depletion" (Walters 1986). In many fisheries, the objective of fisheries management is usually the achievement of the maximum sustainable yield, or *MSY* (Hilborn 2007). However, the maximum economic yield (*MEY*) is a more convenient option, although it is also more difficult to evaluate. Although the basis of studies of the population dynamics of fished species has remained essentially the same, in recent decades fisheries science has made enormous progress relative to traditional views and methods. This is because, since computers became accessible to scientists and fisheries managers for use in their everyday tasks, they have become able to handle enormous amounts of data, which has made it possible to test the outcomes of a wide variety of harvesting scenarios. Unfortunately, many fisheries remain unattended by management efforts or stock assessment studies, meaning that the use of their resources is under free access and they lack harvest controls, and they are thus often depleted (Beddington and Kirkwood 2005). The fisheries of the Caribbean spiny lobster (*Panulirus argus* (Latreille, 1804)) face such challenges, so this chapter focused on analyzing harvest strategies for this species' stocks.

The Caribbean spiny lobster, *Panulirus argus*, belongs to the group of about 20 species of tropical spiny lobsters that are of commercial interest (Holthuis 1991; Phillips et al. 2013). Most details on the life history and fisheries management of this lobster were summarized in the book edited by Phillips (2013). *P. argus* has nocturnal habits, and is associated with coral reef habitats, where it remains hidden in hollows during the day. It is distributed throughout the Caribbean, where it prefers shallow waters but can be found in waters up to 90 m deep or more (Holthuis 1991; Briones-Fourzán and Lozano-Álvarez 2013). It is a gregarious and migratory species. In this region, another species of the same genus (*P. gracilis* (Latreille, 1804)) also occurs, but it is smaller and less abundant than *P. argus* (Holthuis 1991; Briones-Fourzán and Lozano-Álvarez 2013). Lobsters are long-lived species that can live for more than 25 years and grow up to half a meter long, and may reach about 20 kg in weight (Phillips et al. 2013). This species' distribution mainly includes the waters of the Caribbean Sea and the Gulf of Mexico, although some individuals have been found as far north as Massachusetts, northeastern United States of America (USA) (Holthuis 1991; FAO 2019). Towards the south, it can be found until northeastern Brazil (Figure 1), being apparently absent in the waters of the Guianas, where the mouths of several large rivers impose an ecological barrier that may constrain its distribution. It has also been found on the tropical coast of West Africa (Holthuis 1991; FAO 2019).

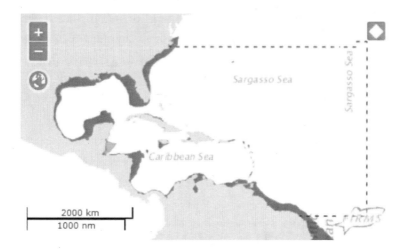

Figure 1. Spiny lobster (*Panulirus argus*) distribution; after the FAO (2019) Fisheries GLOBAL Information System (http://www.fao.org/fishery/species/3445/en).

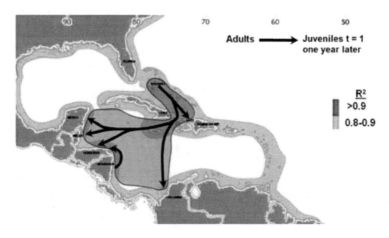

Figure 2. Main pathways of connectivity among spiny lobster populations suggested for the Caribbean waters; after Chávez and Chávez-Hidalgo (2012).

There is some indirect evidence (Grober-Dunsmore and Keller 2008; Butler IV et al. 2010; Chávez and Chávez-Hidalgo 2012) suggesting that currents in the Caribbean and the long larval periods of spiny lobsters are factors that allow there to be high connectivity among the lobster populations of most islands and coastal zones of tropical America (Chávez and Chávez-Hidalgo 2012; Kough et al. 2013) (Figure 2). In the northern part of this species' range, larvae mainly occur in the water column from June to December (FAO 2019).

These lobsters have a life cycle that includes larval stages that live in the plankton for about nine months (Kough et al. 2013; Phillips et al. 2013). This characteristic is one of the reasons for this species' wide distribution. At the end of the larval phase, after passing through a large number of planktonic stages (phyllosomas) and then undergoing metamorphosis, they transform into transparent swimming postlarvae (pueruli) with essentially the same shape as the adult (Briones-Fourzán and Lozano-Álvarez 2013; Phillips et al. 2013). These postlarvae then seek refuge in the reef benthos, where they hide in cavities and grow into juvenile lobsters (Briones-Fourzán and Lozano-Álvarez 2013). In this habitat, they reach sexual maturity after three years, at which point they start breeding (Phillips et al. 2013; FAO 2019).

Spiny lobsters have separate sexes, and although they can reproduce year-round, the main reproductive period takes place from March to September with two main peaks in May and September, corresponding with when water temperatures fall within the range of 23.7 to 28.0°C (Phillips et al. 2013). Females move to deeper water to spawn, and mass migrations occur in the autumn in which groups of animals (up to 50 individuals) move single-file in a certain direction in the daytime, with each animal having body contact with the next in line through the antennae (Phillips et al. 2013; FAO 2019). They lay 800 eggs per g of body weight, with a high fecundity of $> 2 \times 10^{6}$ eggs per ripe female. Spawning takes place with a one-month lag, when the water temperature is generally above 24°C (Pollock 1997; Phillips et al. 2013). During fertilization, the male deposits a spermatophore on the belly of the female, which by its appearance is known as a 'tarspot', from which the eggs are fertilized after spawning (Phillips et al. 2013). After being extruded from the ovary and fertilized, the eggs are adhered to the filaments of the female's pleopods, where they are carried for about one month (Pollock 1997; FAO 2019). At the end of that process, hatching occurs in protected sites on the reef (Briones-Fourzán and Lozano-Álvarez 2013; FAO 2019). Length at settlement, takes place when postlarvae have a size of 6 mm carapace (Pollock, 1997).

Lobsters are predators that mainly feed upon mollusks and crustaceans (trophic level = 3.39 on a scale from 1 to 5, with 5 being top predators) (Phillips et al. 2013). Importantly, in the year 2000, a pathogenic virus was discovered in juvenile Caribbean spiny lobsters from the Florida Keys, USA. *Panulirus argus* is susceptible to infection by *Panulirus argus* virus 1 (PaV1), the only pathogenic virus known to naturally infect any lobster species, which profoundly affects its ecology and physiology. PaV1 is widespread in the Caribbean, with infections reported in Florida (USA), St. Croix, St. Kitts, and the Yucatan, Mexico (Butler IV et al. 2008; Behringer et al. 2012). The sharing of viral alleles among lobsters from distant locations supports the hypothesis of there being high genetic connectivity among lobsters within the Caribbean, and further supports the hypothesis that postlarvae infected with PaV1 may serve to disperse the virus over long distances (Moss et al. 2013). However, *P. argus* lobsters are able to mitigate PaV1 transmission risk by avoiding infected individuals (Behringer et al. 2012).

The Caribbean spiny lobster is a fisheries resource that is highly prized and exploited in all Caribbean countries and Brazil (Holthuis 1991; Phillips et al. 2013; FAO 2019). The most common commercial sizes harvested have a weight of half a kilogram (0.5 kg) and measure about 30 cm in length (Phillips et al. 2013; FAO 2019). In many Caribbean countries, there is open access to lobster harvesting, although some degree of management is applied in most others. For example, in Belize, the following management regulations are in effect: the minimum legal size is set at a carapace length of 3 inches or 7.6 cm or a minimum tail weight of 4 ounces or 113.4 g; there is a closed season from February 5 to June 14. Possession of diced lobster tail meat is prohibited; no commercial deep-water fishing is allowed, nor is lobster fishing with SCUBA gear; no lobster traps are allowed beyond the front reef, nor within a distance of 300 m from any coral formation; and it is prohibited to capture 'berried' or egg-bearing lobsters, lobsters with a 'tarspot', or molting lobsters (FAO 2019). Most Caribbean countries apply management practices analogous to those described above for Belize (Phillips et al. 2013; FAO 2019), although with important differences in the specific rules and regulations applied and their degree of enforcement (Hilborn et al. 2005).

The heterogeneity in management practices among shared or connected lobster stocks (Chávez and Chávez-Hidalgo 2012; Kough et al. 2013; Phillips et al. 2013) and the fact that many are open access (Beddington and Kirkwood 2005) has led to uncertainty regarding the sustainability of and appropriate management approaches for Caribbean lobster fisheries (Pauly et al. 2002). For this reason, the present chapter examined the exploited stocks of the Caribbean spiny lobster (*P. argus*) by addressing some guidelines to improve the criteria of assessing the fisheries for this species in the countries within the Caribbean region in which it is exploited, to ensure that this is done sustainably.

2. MATERIALS AND METHODS

2.1. Data, Models, and Analyses

Exploited Caribbean lobster stocks were assessed using catch data for the last fifteen years available at the Food and Agriculture Organization of the United Nations (FAO, 2011) website (http://www.fao.org/fishery/fishfinder). The values of population parameters came from Ley-Cooper and Chávez (2010). Changes in abundance over time were determined based on changes in the catch data, which were examined in units of metric tons (mt) of fresh weight.

Trends in fishing mortality (F) and estimates of total stock biomass over time were examined. Different fishing scenarios among which the age of first catch (t_c) differed were evaluated based on the F resulting if they were exploited at the level of their maximum sustainable yield (F_{MSY}), as an extreme reference point. The other reference point examined was the F at the maximum economic yield (F_{MEY}). These reference points were determined for each stock at each of the t_c values examined ($1 \geq t_c \leq 6$ years). Tests of t_c values higher than 6 years were deliberately omitted because consumers of this species demand lobsters of around 2 or 3 years old, and the relative value of larger lobsters thus decreases.

The maximum social value of a fishery in a given scenario was determined in two ways. First, the value was assessed in terms of the maximum level of employment or, in other words, the maximum number of fishers, potentially supported by the fishery. The second way in which social value was approached was in terms of the maximum profit per fisher. Economic and social values were assessed in terms of the value per kg landed, the number of fishers per boat, the number of boats, and the number of fishing days, all of which were obtained for the last fishing season in the dataset (FAO, 2011) and then later estimated based on overall trends throughout the 15 years of catch data considered. Costs were estimated by summing the cost/boat/day × the total number of fishing days of the fleet over the fishing season. Profits were subtracted from the total value of the catch minus the total costs. Costs and values were linked to the catch in the semi-automated age-structured FISMO simulation model (Chávez 2005, 2014), which allowed all of the possible scenarios of exploitation to be tested.

The age of first capture was initialized at 3 years and was maintained at a constant value during the fitting process, but then for the simulations all of the F and t_c values feasible to apply to the data were tested to optimize harvest strategies and design exploitation policies for each stock. The age of first maturity of spiny lobsters occurs at the 3-year mark, after the larval and juvenile stages have been completed (Phillips et al. 2013).

Population parameter values were obtained from published sources (Chávez 2009; Ley-Cooper and Chávez 2010). Estimates of the age composition of the catch were made based on FAO (2011) records (http://www.fao.org/fishery/fishfinder). These estimates were made for each one of the main fisheries in the region (those in the Bahamas, Brazil, Cuba, Nicaragua, and the United States of America; Figure 3) over a time scale of 15 years each, and also from a global assessment of the catch data of all 25 spiny lobster fisheries in the Caribbean pooled together. Catch values per kg without value added were obtained as reference data from a small fishery in southeastern Mexico (Ley-Cooper and Chávez 2010), while assuming that their values were approximately the same as those for all other fisheries in the region.

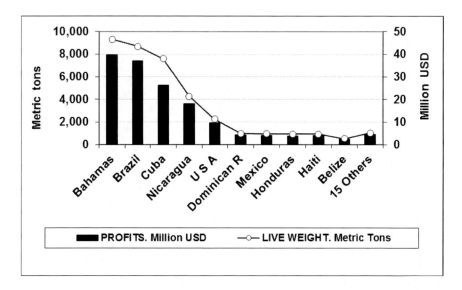

Figure 3. Catch (metric tons) and value (million USD) of the main Caribbean spiny lobster fisheries in the year 2010. Catch data were obtained from FAO (2011) records.

In each case, total mortality (Z_t) was determined with the exponential decay model as follows:

$$N_{a+1} = N_a \cdot e^{(-z_t)} \tag{1}$$

where N_a is the number of spiny lobsters of age a and N_{a+1} is the number of spiny lobsters of age $a+1$ in reconstructed age-groups. The time units used in this equation are years. The von Bertalanffy (1934) growth equation was used to determine the number of lobsters in each age group, as well as their corresponding sizes in terms of their lengths. These lengths were transformed into their respective weights by using the equation:

$$W = 0.0404 \cdot L^3 \tag{2}$$

where W is total weight (g) and L is total length (cm).

The age structures of each age group (in years) were estimated assuming a constant natural mortality rate. For setting the values of the variables in the initial state, the abundance per age class ($N_{a,y}$) was set using the age-specific abundance, $N_a/\Sigma N_a$, obtained from equation (1). In subsequent years, the age structure was defined after the number of one-year-old recruits was estimated. These values were used to calculate the catch-at-age as proposed by Sparre and Venema (1992), and were integrated into the FISMO simulation model as follows:

$$Y_{a,y} = N_{a,y} \cdot W_{a,y} \frac{F_t}{(F_t+M)} \left(1 - e^{-(F_t+M)}\right) \tag{3}$$

where $Y_{a,y}$ is the catch-at-age for age a in each year y, $N_{a,y}$ is the number of spiny lobsters at age a in year y, $W_{a,y}$ is the lobster weight equivalent of $N_{a,y}$, F_t is the fishing mortality (as described earlier in this section), and M is the natural mortality coefficient. Given the initial conditions used, the values of $Y_{a,y}$ were adjusted by varying the initial number of recruits, and then linked to equations (1-3) until the following condition was fulfilled:

$$\Sigma_a^\lambda Y_{a,y} = Y_{y(REC)} \tag{4}$$

where $Y_{y(REC)}$ is the yield recorded during the year y, $a = 3$ years, and $t_\lambda = 3/K$ or the longevity of the species, where K is the growth constant of the von Bertalanffy (1934) growth equation. The value of t_λ was set to 13 years, a value found for this species by assuming that a reasonable life expectancy (L_{max}) is that at which 95% of the population reaches 95% of $L\infty$, the asymptotic length. Thus, the longevity was found by making $L_{max} = 0.95 \times L\infty$ in the von Bertalanffy growth equation and finding the corresponding value of t. The catch equation was used for each year in the time-series analyzed. For the estimation of the natural mortality coefficient (M), the criterion proposed by Jensen (1996, 1997) was adopted, where $M = 1.5K = 0.1793$; the value of K used is further described below. Estimates of the stock biomass and the exploitation rate, $E = [F/(M + F)]$, were made for each age-class in every fishing year analyzed by the model. These values were compared to the E value at the F_{MSY} level. A special case is when E is equal to the F_{MSY} level of fishing mortality, which corresponds to the maximum exploitation rate that a fishery should be able to attain before the stock is overexploited. A diagnosis of the years of the series in which the stock was under- or overexploited was also made, which provided an easy way to recommend either a further increase or decrease in the fishing intensity F.

The annual cohort abundance ($N_{a,y}$) coming from ages older than the age-at-maturity (t_m = 3 years) was used to estimate the annual abundance of adults (S_y) over the years, whereas the abundance of the one-year-old group was used as the number of recruits (R_y). The stock-recruitment relationship was evaluated by using a slightly modified version of the Beverton and Holt (1957) model of the form:

$$R_{y+1} = \frac{a'S_oS_y}{S_y+b'S_o} \tag{5}$$

where R_{y+1} is the number of one-year-old recruits in year $y + 1$, S_y is the number of adults in year y, and S_o is the maximum number of adults in the population. The parameters a' and b' are modified from the original model, such that a' is the maximum number of recruits and b' is the initial slope of the recruitment line, which was kept constant throughout the simulations done herein. The values of the parameters used as input to simulations are shown in Table 1.

Simulations described the main ecological processes underlying the stock dynamics of the spiny lobster fisheries examined. These allowed different exploitation scenarios, with different combinations of fishing intensities and age-at-first catch, to be simulated to see which scenarios maximized the biomass, profits, and social benefits of each fishery. For this purpose, analytical procedures adopting the principles and views of Chávez (1996, 2005) and Grafton et al. (2007) were used. The model used then estimated the simulated catch based on estimates of the stock biomass and fishing mortality for each year of the series for each of the fisheries examined.

Table 1. Population parameter values used for the evaluation of the main spiny lobster fisheries of the Caribbean

Parameter	Value	Units	Model	Source
K	0.24		von Bertalanffy	Chávez (2001)
L_∞	31	Tail length, cm	von Bertalanffy	González-Cano (1991) (mean)
W_∞	1,619	Live weight, g		González-Cano (1991) (mean)
t_o	-0.17	Years	von Bertalanffy	González-Cano (1991) (mean)
a	0.038		Length-weight	Ley-Cooper & Chávez (2010)
b	3.1		Length-weight	Ley-Cooper & Chávez (2010)
t_c	3	Years (both sexes)		Ley-Cooper & Chávez (2010)
t_m	3	Years		Ley-Cooper & Chávez (2010)
t_λ	13	Years		Ley-Cooper & Chávez (2010)
b'	1.77		Beverton and Holt	Ley-Cooper & Chávez (2010)
M	0.36	Instantaneous rate		Ley-Cooper & Chávez (2010)
E_{max}	0.28		$F_{MSY}/(M + F_{MSY})$	This chapter
Value/kg	33	USD		Ley-Cooper & Chávez (2010)
Cost/day/trip	174	USD		Ley-Cooper & Chávez (2010)

Notes: K, t_o, L_∞, and W_∞ are parameters of the von Bertalanffy (1934) growth model; a and b are obtained from the length-weight allometric equation (3); tc = age of first catch; M = natural mortality; t_m = age of maturity; t_λ = longevity; E_{max} = exploitation rate at the *MSY* level. To transform carapace length (*CL*) into total length (*L_t*), the equation used was $L_t = (CL/0.0275) + 3.2$, which gave $L_\infty = 56$ cm $= L_t$, which was the value used as an input in the model. The resulting W_∞ was obtained by using the length-weight equation (3): $W = 0.0404\, L^3$.

The socioeconomics of a fishery were approached through the explicit consideration of the costs of fishing per boat per fishing day. Herein, these values were assessed based on the number of boats, number of fishers per boat, number of fishing days, and the costs of the 2009 fishing season of a small fishery in the southeastern Mexican part of the northwestern Caribbean, while assuming that they were representative of all of the Caribbean fisheries examined. The value was set to the selling price at the dock of the spiny lobsters landed in the same fishery; the profit was the difference between the costs and the value. In these simulations, the costs of fishing and catch value per kg were assumed to be constant over time.

The socioeconomic information used allowed the economic trends within each of the fisheries analyzed to be reconstructed over the 15 years analyzed with the aid of the simulation model. This was done by using the estimated fishing mortality for each fishery over time as a reference value to estimate their corresponding value of each of the different economic variables.

Changes in each population's lobster biomass were estimated using the number of survivors in each cohort. The estimation of the potential yield in each case allowed the *MSY* and the *MEY* values for each t_c (where $1 \leq t_c \leq 6$ years) to be found.

2.2. The Simulations

The estimated catch values showed that the level of exploitation at which the rate of change in the catch or profits of a fishery with respect to changes in its *F* was zero corresponded to the point at which the maximum values of these endpoints were attained, which was equivalent to the *MSY* of the catch and the *MEY* of the profits, respectively. A parallel trend in the *F* and t_c values was required for the *MSY* and *MEY* to be found in most cases. The uncertainties of the estimates produced, after Hilborn and Liermann (1998), were expressed as coefficients of variation, estimated for each fishery, but for simplicity their values were deliberately ignored in the presentation of results, and instead the mean tendencies were presented.

Parameter values were fixed at some 'best' estimated value herein rather than allowing for uncertainty in their values (Hilborn and Liermann 1998); simulations were done in this way to allow clear trends to be obtained and allow specific harvest recommendations to be addressed.

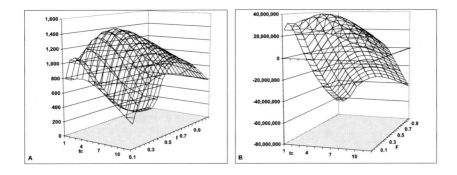

Figure 4. The response of a stock's yield (A) and value (B) to differing values of F and t_c. Simulation outputs for the Florida spiny lobster fishery (in 2007) are presented. Potential yield (A) is presented in metric tons, and potential profits (B) are in USD.

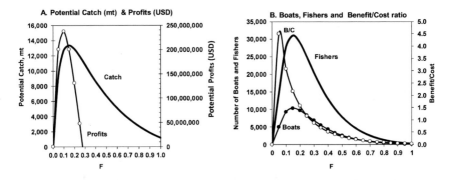

Figure 5. Simulation outputs of potential yield (mt) (A) and potential profits (USD) (B) of the Florida spiny lobster fishery at different levels of F and a constant $t_c = 3$ years. The *MSY* and *MEY* values are the peaks of the curves for catch and profit, respectively. The *MSY* value corresponds to the catch in the years 1997 and 1999.

2.3. Theoretical Statement

Results of previous stock assessments indicated that yield displays a dome-shaped response surface as a function of F and t_c, as shown in Figure 4A and B. When a single set of outputs as a function of F was examined at a certain t_c, the yield and profits also displayed curves that peaked and attained their maximum values at the corresponding *MSY* and *MEY* (Figure 5A, B). In general, these peak values were found at the same t_c, but the *MSY* was usually attained at a higher F than the *MEY* was. In highly valued fisheries like those of the spiny lobster, the F resulting in the *MEY* generally coincides with that of the *MSY*. Additionally, there should be a certain F value where the maximum yield is attained, declining at higher as well as lower F values, such as in the well-known figure of yield isopleths presented by Beverton and Holt (1957) that displayed the yield per recruit as a function of F and t_c.

3. RESULTS

3.1. Optimum Yields

As result of the examination of the long-term stock responses when attempting to maximize yields, it was found that this variable tended to increase with higher values of t_c. In some cases, the yield at a $t_c \leq 5$ was higher by 20% than that at the current t_c of 3 years, as shown in Figure 6A-C. In three of the fisheries examined (Nicaragua, the Bahamas, and Cuba), the F values required to reach the *MSY* were higher than those needed to reach the *MEY*. In the other fisheries, the F levels at the *MSY* and *MEY* overlapped.

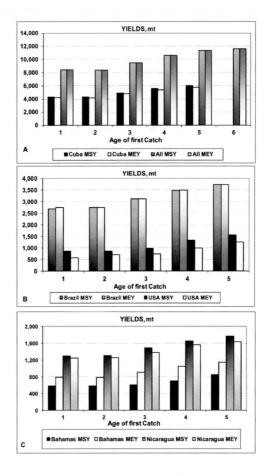

Figure 6. Potential *MSY* and *MEY* of representative Caribbean spiny lobster fisheries as a function of t_c (x-axes) and F (not displayed here). Results are presented for all 25 Caribbean fisheries and Cuba (A), for Brazil and the USA (B), and for Nicaragua and the Bahamas (C). In most cases, the yield at the *MSY* was higher than that at the *MEY*, and yields were higher with a higher t_c.

3.2. Optimum Economic Yields

The profits of these fisheries were higher at the *F* values applied to reach the *MEY* than at those required for the *MSY*, as shown in Figure 7A-C. The *MEY* tended to increase as a function of the t_c in the cases of Nicaragua and Brazil. In Cuba and the USA, this increasing trend was not very significant, and higher values were obtained when the *F* of the *MEY* was used rather than that of the *MSY*. These trends depended on the volume of spiny lobsters landed. The case of the Bahamas was remarkable, because the yields tended to decrease at all t_c values examined, and the profits were negative when the *F* values required for the *MSY* were applied (Figure 7B). The case of the USA was not so critical because the line of profits was around the economic equilibrium level for this country, with profits being negative at only $t_c = 4$ years (Figure 7C). Evidently, operating costs are sensitive to the size of the exploited stocks (Hannesson 2007).

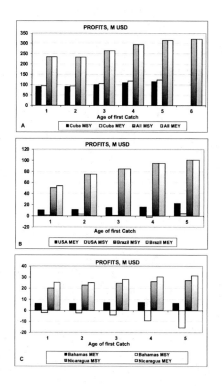

Figure 7. Potential profits (in millions of USD) of representative Caribbean spiny lobster fisheries at the *MSY* and *MEY* as a function of t_c and *F*. Results are shown for all 25 Caribbean fisheries combined and Cuba (A), for Brazil and the USA (B), and for Nicaragua and the Bahamas (C). In most cases, the yield at the *MEY* was higher than that at the *MSY*. In addition, profits were generally higher with a higher t_c, except in the Bahamas, where the fishery was not profitable in any scenario when the *F* applied reached the *MSY*.

3.3. Social Benefits

As described earlier (section 2.1), in the simulation model the economic and social values were linked to the costs and value of fishing activities, as well as to the number of fishers involved directly in this activity. Therefore, the model is sensitive to fishing effort and the resulting numbers of fishers, as shown in Figure 8A and B. The same increasing trend was observed for the number of fishers as was previously seen in the case of yield (Figure 7A-C), wherein a higher number of fishermen was required when t_c values were higher as a consequence of the higher catches under these conditions. It was particularly remarkable to observe the case in which all of the fisheries were grouped together (Figure 8B), wherein nearly 30,000 fishers currently participate in all of the Caribbean lobster fisheries with a $t_c = 3$ years, but this could be increased to nearly 50,000 if t_c values were raised to 5 or 6 years. The Brazilian fishery displayed the steepest slope of this growing trend in relation to t_c (Figure 8A). In a similar way as in some of the results already described, when the catch was set at the *MSY*, the number of direct jobs potentially supported by the fishery was the same or higher than that at the *MEY*.

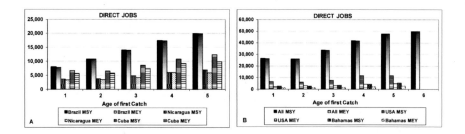

Figure 8. Number of fishers involved in the main Caribbean lobster fisheries and in all of the fisheries combined in relation to t_c. Increases were observed when higher t_c values were used. The catch rate per fisher was maintained at a constant value, representing the mean value under current conditions. Results are shown for Brazil, Cuba, and Nicaragua (A) and for the USA, the Bahamas, and all 25 fisheries combined (B).

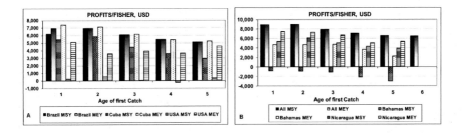

Figure 9. Social benefits, expressed as profits per fisher, of the five main Caribbean spiny lobster fisheries and all of the Caribbean fisheries combined. Results are shown for Brazil, Cuba, and the USA (A) and for the Bahamas, Nicaragua, and all 25 Caribbean fisheries combined (B). The social benefit mainly declined with increasing t_c, and the social benefit was often much higher at the *MEY* than at the *MSY*.

The social benefit of a fishery was expressed herein as the profits obtained per fisher, which was simply calculated by dividing the profits by the number of fishers while assuming that they would have access to the same catch as in current conditions. In all cases except one, the social benefits of the fisheries declined when the stocks were exploited with higher t_c values (Figure 9A, B). The Bahamas fishery was not profitable when the *F* values required for the *MSY* were applied. In the USA, the social benefits were much higher under the *MEY* scenario.

4. DISCUSSION

4.1. The Model Output

To begin, it must be remembered that the figure presented by Beverton and Holt (1957) displaying the relationship of the yield-per-recruit to *F* and t_c did not explicitly take recruitment into account. When a simulation is used as in the present case (Chavez 2005, 2014), a recruitment function must be explicitly defined, as this is a necessary condition to link the simulated cohorts over time. In earlier sections (2.1, 2.2, and 2.3), the simulations were described as being expected to provide the same kind of performance by estimating the stock biomass as the yield-per-recruit curve at high F and t_c values. However, this hypothesis has to be rejected. Therefore, it is easy to understand why yields and profits were found to increase when high t_c values were considered because, in this case, the size of the breeding stock would be higher than that under current conditions, where there is a general tendency to catch smaller animals as a consequence of increasing competition among fishers. It must also be kept in mind that there should be a maximum number of adults and recruits that can occur within a stock, which is a limit imposed by the carrying capacity of the ecosystem. For example, in the case of adult biomass, the carrying capacity may be close to the virgin biomass of an unexploited stock, or, in other words, it could be twice the stock biomass at the *MSY* level.

In several instances, the approach presented herein was consistent with the methods used in the management strategy evaluation (MSE) approach, which relies on testing the whole management process through simulations using and specifying different performance measures with which to compare alternative management strategies (Sainsbury et al. 2000). In this chapter, 15-year time series of catch data were chosen to feed the model for two reasons: one was to minimize the effects of climate variability, which may induce significant changes in recruitment rates if longer time series are used; and the other was to be consistent with Jackson et al. (2001), who stated that "Retrospective data not only help to clarify underlying causes and rates of ecological change", but also help to "demonstrate achievable goals for restoration and management."

The estimation of potential yield is one of the goals to which any fishery's stock assessment should aspire, even if it is based on simple relations (Beddington and Kirkwood 2010). In the Introduction (section 1), it was mentioned that the *MEY* is a more convenient option for a target of fisheries assessment than the *MSY* because it is often achieved at lower levels of F, thus reducing the risks of overexploiting the stock. This may be a function of the discount rate (Grafton et al. 2007). In addition, in the method applied herein to assess stock biomass, Hilborn and Walters (1992) can be quoted in saying that "you cannot find the top of a [yield] curve without going beyond the top". The "top" they mention is the *MSY* level. In contrast, when the *MEY* is adopted as a target, when going beyond the top of the *MEY* curve the stock is still underexploited, or at least has not exceeded its reproductive potential. This was the case for about half of the Caribbean spiny lobster fisheries examined in this chapter.

4.2. Management Options

The Caribbean spiny lobster fisheries are a clear example of unsuccessful management systems, wherein access to the resource is open and there is poor ability to monitor its exploitation and implement regulations (Hilborn et al. 2005), and consequently these fisheries have rarely been sustainable (Pauly et al. 2002).

We are living in a world where the political incentive is to maintain the continuous exploitation of fished stocks, which has led to the generally depleted condition of fish stocks worldwide (Rosenberg 2003). Therefore, the purpose of the exercises done in this chapter was to provide some guidelines that could help achieve the sustainable management of Caribbean spiny lobster fisheries (Mora et al. 2009). The fisheries examined herein displayed great heterogeneity. For this reason, it was not possible to make a few general management recommendations that could be feasibly adopted for widespread application. However, based on the results, there are at least three reference points that can potentially be used to develop some planning guidelines, which are the F_{MSY}, F_{MEY}, and t_c. Further, the yield, profits, and social benefit (herein referring to the profits per fisher) are the dependent variables that should be compared when assessing the performance of a fishery under different conditions. The combination of the F and the t_c that produces the highest catch, profits, and social benefit is the feasible targets that should be sought for the management of these fisheries. In this approach, the economic and social aspects of the fisheries were incorporated, which agrees with modern trends in the definition of sustainability (Quinn II and Collie 2005), wherein the costs of management and the windfall profits of fisheries conservation should be shared (Hilborn et al. 2005).

As a final corollary, and after perceiving that the spiny lobster stocks are exploited under heterogeneous conditions and different fishing gears, it is apparent that recurrent socio-economic crises in the exploitation of this resource can and do take place in some countries within the species' distribution. Therefore, it would be useful to create an international organization in charge of regulating and controlling regional access to this resource. This group would be in charge of carrying out stock assessments of the main stocks and assign catch quotas and minimum size limits every year, before the beginning of the next fishing season. In this way, the sustainable exploitation of this common property resource would be ensured. The adoption of these statements as general principles in stock assessment and management is therefore suggested, as well as those more specific principles recommended in a previous paper (Chávez 2007).

CONCLUSION

As result of the analyses carried out in this chapter, the following conclusions were derived:

1. The *MSY* is attained at the same or higher F values as those required to reach the *MEY*. The highest values for these variables are obtained when $t_c \geq 5$ years.
2. Likewise, the maximum profits are reached with the F values required to achieve the *MEY*. In some cases, when the F required for the *MSY* to be reached is applied, the fishery may become non-profitable.
3. The maximum social benefit, meaning the number of jobs directly supported by the fishery, is obtained at the F values required to reach the *MSY* and when $t_c \geq 5$ years.
4. The maximum economic benefit per fisher is obtained when the F required to attain the *MEY* is applied, and when $t_c \leq 3$ years.

ACKNOWLEDGMENTS

Rene Buesa kindly provided catch data for the Florida spiny lobster fishery, read the manuscript, and made valuable suggestions. The author holds a research fellowship from COFAA and EDI, IPN.

REFERENCES

Beddington J, Kirkwood GP (2005) Introduction: fisheries, past, present and future. *Phil Trans R Soc B* 360:3-4.

Beddington JR, Kirkwood GP (2010) The estimation of potential yield and stock status using life-history parameters. *Phil Trans R Soc B* 360:163-170.

Behringer DC, Butler MJ IV, Stentiford GD (2012) Disease effects on lobster fisheries, ecology, and culture: overview of DAO Special 6. *Dis Aquat Organ* 100:89-93.

Beverton RJ, Holt SS (1957) On the dynamics of exploited fish populations. *UK Min Agric Fish Inv Ser 2* 19:1-533.

Briones-Fourzán P, Lozano-Álvarez E (2013) Essential habitats for Panulirus spiny lobsters. In: Phillips BF (ed.) *Lobsters: Biology, Management, Aquaculture and Fisheries* (2nd Ed.). Oxford, UK: John Wiley and Sons Ltd., p. 186-220.

Butler MJ IV, Behringer DC, Shields JD (2008) Transmission of *Panulirus argus* virus 1 (PaV1) and its effect on the survival of juvenile Caribbean spiny lobster. *Dis Aquat Organ* 79:173-182.

Butler MJ IV, Mojica AM, Sosa-Cordero E, Millet M, Sanchez-Navarro P, Maldonado MA, Posada J, Rodriguez B, Rivas CM, Oviedo A, Arrone M, Prada M, Bach N, Jimenez N, Garcia-Rivas MdC, Forman K, Behringer DC, Matthews T, Paris C, Cowen R (2010) Patterns of spiny lobster (*Panulirus argus*) postlarval recruitment in the Caribbean: a CRTR project. *Proc Gulf Caribbean Fish Inst* 62:361-369.

Chávez EA (1996) Simulating fisheries for the assessment of optimum harvesting strategies. *Naga ICLARM* 19:33-35.

Chávez EA (2001) Policy design for spiny lobster (*Panulirus argus*) management at the Meso-American Barrier Reef System. *Crustaceana* 74:1119-1137.

Chávez EA (2005) FISMO: A generalized fisheries simulation model. In: Kruse GH, Gallucci VF, Hay DE, Perry RI, Peterman RM, Shirley TC, Spencer PD, Wilson B, Woodby D (eds.) *Fisheries Assessment and Management in Data-Limited Situations*. Fairbanks, AK, USA: Alaska Sea Grant College Program, University of Alaska Fairbanks, p. 659-681.

Chávez EA (2007) Socio-economic assessment for the management of the Caribbean spiny lobster. *Proc Gulf Caribbean Fish Inst* 60:193-196.

Chávez EA (2009) Potential production of the Caribbean spiny lobster (Decapoda, Palinura) fisheries. *Crustaceana* 82:1393-1412.

Chávez EA (2014) Un modelo numérico para la administración sustentable de las pesquerías. [A numerical model for the sustainable management of Fisheries]. *CICIMAR Oceánides* 29:45-56. [In Spanish].

Chávez EA, Chávez-Hidalgo A (2012) Pathways of connectivity amongst the western Caribbean spiny lobster stocks. *12th Int Coral Reef Symp*.

Cushing DH (1968) *Fisheries Biology: A Study in Population Dynamics*. Madison, WI: University of Wisconsin Press, 200 p.

FAO – Food and Agriculture Organization of the United Nations (2011) FAO FishFinder. Institutional Websites. *FAO Fisheries & Aquaculture Department* [online]. Rome, Italy: Food and Agriculture Organization of the United Nations (FAO/UN). [Updated 11 December 2011]. Available at: http://www.fao.org/fishery/fishfinder/.

FAO (2019) *Panulirus argus* (Latreille, 1804). Species Fact Sheets. *FAO Fisheries & Aquaculture Department* [online]. Rome, Italy: Food and Agriculture Organization of the United Nations (FAO/UN). [Accessed 13 February 2019]. Available at: http://www.fao.org/fishery/species/3445/en.

González-Cano JM (1991) *Migration and refuge in the assessment and management of spiny lobster* Panulirus argus *in the Mexican Caribbean*. Ph.D. thesis, Imperial College London, London, UK.

Grafton RQ, Kompas T, Hilborn RW (2007) Economics of overexploitation revisited. *Science* 318:7.

Grober-Dunsmore R, Keller BD (eds.) (2008) *Caribbean Connectivity: Implications for Marine Protected Area Management. Proceedings of a Special Symposium, 9-11 November 2006, 59ᵗʰ Gulf and Caribbean Fisheries Institute, Belize City, Belize*. Marine Sanctuaries Conservation Series ONMS-08-07, 195 p.

Gulland JA (1969) Manual of methods for fish stock assessment. Part I: Fish population analysis. *FAO Man Fish Sci* 4:1-154.

Gulland JA (ed.) (1988) *Fish Population Dynamics: The Implications for Management*. Chichester, UK, Wiley, 422 p.

Hannesson R (2007) A note on the "stock effect". *Mar Resour Econ* 22:69-75.

Hilborn R (2007) Defining success in fisheries and conflicts in objectives. *Mar Policy* 31:153-158.

Hilborn R, Liermann M (1998) Standing on the shoulders of giants: learning from experience in fisheries. *Rev Fish Biol Fisher* 8:273-283.

Hilborn R, Parrish JK, Little K (2005) Fishing rights or fishing wrongs? *Rev Fish Biol Fisher* 15:191–199.

Hilborn R, Walters CJ (1992) *Quantitative Fisheries Stock Assessment: Choice, Dynamics and Uncertainty*. New York, NY, USA: Chapman & Hall, 570 p.

Holthuis LB (1991) *Marine Lobsters of the World. An Annotated and Illustrated Catalogue of the Species of Interest to Fisheries Known to Date*. FAO Species Catalogue No. 125, Vol. 13. Rome, Italy: Food and Agriculture Organization of the United Nations (FAO/UN), 292 p.

Jackson JBC, Kirby MX, Berger WH, Bjorndal KA, Botsford LW, Bourque BJ, Bradbury RH, Cooke R, Erlandson J, Estes JA, Hughes TP, Kidwell S, Lange CB, Lenihan HS, Pandolfi JM, Peterson CH, Steneck RS, Tegner MJ, Warner RR (2001) Historical overfishing and the recent collapse of coastal ecosystems. *Science* 293:629-638.

Jensen AM (1996) Beverton and Holt life history invariants result from optimal trade off of reproduction and survival. *Can J Fish Aquat Sci* 53:820-822.

Jensen AM (1997) Origin of the relation between K and Linf and synthesis of relations among life history parameters. *Can J Fish Aquat Sci* 54:987-989.

Kough AS, Paris CB, Butler MJ IV (2013) Larval connectivity and the international management of fisheries. *PLoS ONE* 8:e64970. DOI:10.1371/journal.pone.0064970.

Latreille PA (1804) Des langoustes du muséum national d'histoire naturelle. [The lobsters of the national museum of natural history]. *Ann Mus Hist Nat Paris* 3:388-395. [In French].

Ley-Cooper K, Chávez EA (2010) Bio-economic modelling applied to a spiny lobster fishery of the northwestern Caribbean. *Proc Gulf Caribbean Fish Inst* 62:148-159.

Mora C, Myers RA, Coll M, Libralato S, Pitcher TJ, Sumaila RU, Zeller D, Watson R, Gaston KJ, Worm B (2009) Management effectiveness of the world's marine fisheries. *PLoS Biol* 7(6):e1000131. DOI:10.1371/journal.pbio.1000131.

Moss J, Behringer D, Shields JD, Baeza A, Aguilar-Perera A, Bush PG, Dromer C, Herrera-Moreno A, Gittens L, Matthews TR, McCord MR, Schärer MT, Reynal L, Truelove N, Butler MJ IV (2013) Distribution, prevalence, and genetic analysis of *Panulirus argus* virus 1 (PaV1) from the Caribbean Sea. *Dis Aquat Organ* 104:129-140.

Pauly D, Christensen V, Guénette S, Pitcher TJ, Sumaila UR, Walters CJ, Watson R, Zeller D (2002) Towards sustainability in world fisheries. *Nature* 418:689-695.

Phillips BF (ed.) (2013) *Lobsters: Biology, Management, Aquaculture and Fisheries* (2nd Ed.). Oxford, UK: John Wiley and Sons Ltd.

Phillips BF, Melville-Smith R, Kay MC, Vega-Velázques A (2013) *Panulirus* species. In: Phillips BF (ed.) *Lobsters: Biology, Management, Aquaculture and Fisheries* (2nd Ed.). Oxford, UK: John Wiley and Sons Ltd., p. 289-325.

Pollock D.E. 1997. Egg production and life-history strategies in some clawed spiny lobster populations. Bull. Mar. Sci., 61(1):97-107.

Quinn TJ II, Collie JS (2005) Sustainability in single-species population models. *Phil Trans R Soc B* 360:147-162.

Quinn TJ II, Deriso R (1999) *Quantitative Fish Dynamics.* Oxford, UK: Oxford University Press, 542 p.

Ricker WE (1975) Computation and interpretation of biological statistics of fish populations. *Bull Fish Res Board Can* 191:1-382.

Rosenberg A (2003) Managing to the margins: the overexploitation of fisheries. *Front Ecol Environ* 1:102-106.

Sainsbury KJ, Punt AE, Smith ADM (2000) Design of operational management strategies for achieving fishery ecosystem objectives. *ICES J Mar Sci* 57:731-741.

Sparre P, Venema, S (1992) Introduction to tropical fish stock assessment. Part 1 manual. *FAO Fish Tech Paper* 306/1:1-376.

von Bertalanffy L (1934) Untersuchungen uber die gesetzlichkeiten des wachstums. 1. Allgemeine grundlagen der theorie. [Investigations on the laws of growth. 1. General principles of theory]. *Roux'Arch. Entwicklungsmech Org* 131:613-653. [In German].

Walters CJ (1986) *Adaptive Management of Renewable Resources.* Basingstoke, NY, USA: MacMillan Publishers Ltd., 374 pp.

ABOUT THE EDITOR

Brady K. Quinn is a researcher at the University of New Brunswick (Saint John Campus) who has been involved in numerous research projects on the thermal biology, behavior, development, conservation, fisheries ecology, and physiology of crustaceans in general, and lobsters in particular. His work with American lobster larvae has focused on their temperature-dependent development and dispersal, and considerations of the implications thereof to stock connectivity and fisheries management. He is also an active teacher of biological science and participant in marine science outreach, and now also works as a scientific English language editor with Cactus Communications (Editage).

INDEX

A

adaptation, 9, 40, 62, 63, 106
American lobster, v, x, 1, 2, 3, 4, 5, 8, 9, 11, 12, 14, 15, 16, 18, 20, 21, 23, 26, 27, 30, 32, 35, 39, 40, 42, 44, 48, 49, 50, 51, 52, 53, 54, 55, 56, 57, 59, 62, 63, 104, 105, 106, 108, 109, 110, 111, 113, 116, 117, 149
anatomy, 107
aquaculture, 113
arachnids, 6, 44, 49
arthropods, 43, 53, 57
Astacus astacus, 63, 66, 68, 70, 71, 72
Athanas nitescens, 64, 69

B

behavior, ix, xi, 62, 63, 65, 71, 84, 89, 93, 94, 96, 97, 98, 99, 100, 105, 106, 110, 118
benefits, 121, 131, 139
biodiversity, 62, 97, 101, 118, 119
biogeography, 108, 117
biologically relevant, 5, 15, 23, 49

biology, viii, ix, 52, 53, 54, 55, 57, 59, 62, 105, 106, 107, 108, 109, 110, 111, 114, 115, 116, 117, 118, 119, 143, 144, 146, 149
biomass, 127, 130, 131, 133, 139, 140
biotope, x, xi, 61, 63, 70, 77, 89, 92, 97, 99
body shape, vii, 63, 64, 66, 77, 95
bootstrap confidence intervals, 3, 8, 12, 13, 14
bootstrapping, 14, 15
breeding, 125, 140

C

Cancer pagurus, 63, 66, 68, 69, 70, 71, 72, 76, 79, 89, 94, 105, 110, 120
carapace, 64, 66, 67, 73, 74, 77, 78, 81, 95, 100, 126, 132
Carcinus maenas, 63, 66, 68, 69, 70, 71, 72, 76, 79, 89, 94, 107, 108, 111
Caribbean, v, ix, xii, 59, 121, 122, 123, 124, 126, 127, 128, 129, 132, 133, 136, 137, 138, 139, 141, 143, 144, 145, 146
Caribbean countries, 121, 126
Caribbean Sea, 123, 146

Caribbean spiny lobster, ix, xii, 59, 123, 126, 127, 129, 136, 137, 139, 141, 143, 144

catch, ix, xii, 102, 111, 122, 127, 128, 130, 131, 132, 133, 134, 138, 139, 140, 141, 142, 143

chelae, 64, 66, 73, 74, 76, 77, 93, 98, 99, 100

climate, viii, ix, x, 2, 4, 5, 9, 43, 45, 50, 51, 53, 57, 58, 103, 114, 140

climate change, viii, ix, x, 2, 4, 5, 9, 45, 50, 51, 53, 57

clustering, 65, 73, 74, 75, 82, 91, 92

composition, 99, 109, 128

conservation, 56, 83, 97, 108, 119, 122, 141, 149

crabs, vii, xi, 63, 77, 79, 80, 88, 89, 95, 96, 100, 107, 109, 110, 114, 115, 116

D

decapod crustaceans, vii, 44, 49, 51, 61, 62, 63, 82

development, v, ix, 1, 2, 3, 5, 6, 7, 8, 9, 10, 11, 12, 14, 15, 16, 18, 20, 21, 23, 26, 27, 30, 32, 33, 34, 35, 36, 39, 40, 41, 42, 43, 46, 47, 48, 49, 50, 51, 52, 53, 54, 55, 56, 57, 58, 59, 80, 82, 99, 111, 114, 120, 149

development rate, ix, 2, 3, 5, 7, 8, 12, 16, 18, 20, 21, 23, 26, 30, 39, 48, 54, 57

development time, 5, 7, 9, 10, 11, 12, 13, 16, 49, 52, 57

diet, 65, 71, 79, 107

distribution, 5, 14, 58, 68, 99, 103, 106, 110, 122, 123, 124, 125, 142

diversity, xi, 62, 65, 70, 73, 74, 86, 87, 92, 98

E

ecological processes, 118, 131

ecology, ix, 3, 9, 53, 55, 56, 57, 62, 64, 85, 95, 101, 106, 107, 108, 109, 111, 113, 115, 118, 126, 143, 149

ecomorphology, 62, 77, 106

ecosystem, 114, 118, 119, 140, 147

ectothermic, 3, 55

edible crab, 63, 110

egg, 54, 79, 80, 102, 127

environment, 55, 95, 112

European lobster, x, 58, 61, 62, 63, 89, 92, 102, 104, 105, 107, 110, 111, 112, 113, 114, 115, 116, 117

European spiny lobster, 63, 106, 107, 109, 110

evolution, 56, 63, 82, 118

F

fauna, 112, 114

fecundity, 63, 65, 68, 80, 82, 115, 125

fish, 90, 100, 101, 115, 141, 143, 145, 147

fisheries, viii, ix, x, xii, 2, 4, 45, 50, 51, 57, 59, 102, 108, 109, 110, 118, 120, 121, 122, 123, 124, 126, 127, 128, 129, 131, 132, 133, 134, 135, 136, 137, 138, 139, 140, 141, 142, 143, 144, 145, 146, 147, 149

fishing, vii, viii, xii, 87, 114, 119, 126, 127, 128, 130, 131, 133, 138, 142, 145

FISMO simulation model, 128, 130

fitness, 52, 58

flexibility, 64, 66

food, 8, 41, 59, 78, 87, 88, 90, 116

freshwater, viii, 115

G

Galathea strigosa, 63, 66, 68, 69, 70, 71, 72, 76, 89, 94

growth, 7, 16, 49, 53, 54, 56, 59, 82, 83, 88, 99, 106, 107, 110, 111, 115, 116, 129, 130, 132, 147
growth rate, 53, 54, 107

H

habitat, viii, ix, x, 3, 44, 62, 63, 64, 65, 70, 77, 78, 82, 83, 84, 85, 87, 91, 92, 95, 97, 98, 99, 100, 101, 103, 107, 108, 112, 114, 117, 123, 125, 143
harvesting, xii, 121, 123, 126, 144
Homarus americanus, viii, 2, 4, 52, 53, 54, 55, 56, 57, 59, 62, 63, 66, 68, 69, 70, 71, 72, 94, 105, 106, 107, 108, 109, 110, 111, 113, 114, 115, 116
Homarus gammarus, viii, 54, 58, 61, 63, 66, 68, 69, 70, 71, 72, 76, 79, 88, 89, 91, 94, 104, 105, 106, 107, 109, 110, 111, 112, 113, 114, 115, 116, 117

I

insects, 6, 8, 44, 49, 55, 58
intrinsic optimum temperature, v, ix, 1, 2, 3, 7, 8, 9, 14, 33, 35, 40, 44, 55, 58

L

laboratory studies, xi, 47, 62, 64, 77, 83, 92, 97, 101
larvae, ix, x, xi, 2, 3, 4, 5, 6, 8, 9, 10, 12, 14, 15, 16, 18, 20, 21, 23, 26, 27, 30, 31, 33, 35, 39, 40, 41, 42, 43, 44, 45, 49, 50, 51, 52, 54, 56, 57, 59, 62, 80, 82, 85, 97, 102, 116, 125, 149
larval development, x, 2, 3, 5, 6, 7, 8, 9, 10, 14, 15, 16, 33, 34, 40, 46, 48, 49, 51, 52, 59

larval stages, x, 2, 4, 6, 9, 13, 14, 15, 33, 34, 43, 44, 56, 64, 66, 80, 82, 85, 125
life cycle, viii, ix, x, 56, 92, 125
life history, xi, 53, 62, 63, 65, 68, 80, 98, 99, 101, 106, 107, 117, 123, 146
longevity, 115, 117, 130, 132

M

marine fish, 107, 146
marine species, 80, 98, 111
maturation, 65, 68, 80
metamorphosis, 82, 102, 103, 114, 125
microhabitats, 101, 103
migration, 56, 71, 73, 74, 93
morphology, 62, 63, 64, 77, 95, 98, 99, 100, 105, 107, 108
mortality, xii, 3, 45, 56, 122, 127, 129, 130, 132, 133

N

Nephrops norvegicus, viii, 63, 66, 68, 69, 70, 71, 72, 76, 79, 89, 116
North Atlantic Ocean, 62
northern deep-water shrimp, 64
Norway lobster, 63, 116

P

Palinurus elephas, 63, 66, 68, 70, 71, 72, 80, 105, 106, 107, 108, 109, 110
Pandalus borealis, 64, 66, 68, 69, 70, 71, 72, 100, 109, 113, 115
Panulirus argus, ix, 59, 122, 123, 124, 126, 143, 144, 145, 146
Panulirus argus virus 1 (PaV1), 126, 143, 146
Paralithodes camtschaticus, 64, 66, 68, 69, 70, 71, 72, 76, 79, 94

pereiopods, 64, 66, 76, 77, 95, 98, 100
pigmentation, 63, 65, 66, 78, 92, 95, 98, 99, 100, 107
plankton, xi, 10, 59, 62, 85, 102, 125
predation, 3, 64, 65, 85, 92, 98, 105, 108, 114, 117
predator, ix, xi, 4, 45, 62, 65, 70, 79, 81, 82, 83, 84, 87, 90, 92, 94, 95, 98, 100, 101, 102, 103, 106, 112, 115, 126
protected areas, 103, 113, 116
protection, 83, 95, 97, 98, 99, 100

R

red king crab, 64, 78, 107
regression, 12, 46, 48, 58
reproduction, 80, 82, 104, 114, 146

S

salinity, 56, 59, 116
sediment, viii, 70, 95, 100
settlement, 13, 45, 50, 52, 54, 55, 59, 63, 78, 84, 85, 86, 87, 88, 90, 92, 102, 126
Sharpe-Schoolfield-Ikemoto (SSI) model, ix, x, 2, 3, 6, 7, 8, 9, 10, 11, 12, 13, 14, 15, 16, 17, 18, 19, 20, 21, 22, 23, 24, 25, 26, 28, 29, 30, 31, 32, 35, 36, 37, 38, 39, 40, 41, 42, 43, 44, 46, 49, 46, 47, 49, 50, 51, 52, 55, 58
shelter, x, xi, 65, 70, 85, 88, 89, 90, 94, 95, 96, 98, 99, 101, 102, 103, 106, 108, 114, 116
shoreline, xi, 62
shrimp, 64, 78, 88, 104, 109, 115
size, ix, 16, 63, 65, 68, 73, 78, 80, 82, 85, 88, 93, 97, 99, 100, 106, 108, 109, 114, 126, 137, 140, 142
social behavior, 65, 71
social benefits, 131, 139
social value, xii, 121, 128, 138

socioeconomic, ix, xii, 121, 133
species, viii, ix, x, xi, 2, 3, 4, 5, 6, 7, 8, 9, 10, 40, 41, 43, 44, 45, 46, 48, 49, 50, 51, 55, 56, 57, 58, 61, 62, 63, 64, 65, 66, 68, 69, 70, 71, 72, 73, 74, 75, 76, 77, 78, 80, 82, 83, 84, 87, 88, 89, 90, 91, 92, 93, 95, 96, 97, 98, 101, 103, 109, 111, 115, 122, 123, 124, 125, 126, 127, 128, 130, 142, 144, 146, 147
spiny lobster, v, ix, xii, 59, 63, 78, 105, 106, 107, 108, 109, 110, 121, 122, 123, 124, 126, 127, 128, 129, 130, 131, 132, 133, 134, 135, 136, 137, 139, 141, 142, 143, 144, 145, 146, 147
spiny squat lobster, 63
stock, xii, 97, 102, 104, 110, 123, 127, 128, 130, 131, 134, 135, 139, 140, 142, 143, 145, 147, 149
stock assessment, 123, 135, 140, 142, 145, 147
stock biomass, 127, 130, 131, 139, 140
survival, ix, x, xi, 2, 5, 7, 8, 13, 47, 49, 54, 56, 57, 59, 62, 65, 78, 80, 83, 88, 92, 102, 105, 107, 110, 116, 117, 143, 146
survivors, 133
sustainability, ix, xii, 119, 127, 141, 146
sustainable, xii, 122, 127, 141, 142, 144

T

tail, vii, 64, 65, 66, 67, 71, 73, 74, 77, 78, 93, 95, 100, 126
temperature, viii, ix, 2, 3, 4, 5, 6, 7, 8, 9, 10, 12, 13, 14, 16, 26, 33, 35, 40, 41, 43, 44, 45, 47, 48, 50, 52, 53, 54, 55, 56, 57, 58, 59, 72, 73, 74, 81, 98, 116, 125, 149
temperature dependence, 52
temporal variation, 59
thermodynamics, 2, 3, 6, 8
threshold temperature, ix, 3, 6, 7, 8, 9, 34

W

water, ix, x, xi, 4, 40, 44, 45, 51, 57, 62, 64, 77, 78, 85, 89, 100, 101, 113, 125, 126
worldwide, 115, 141

Y

yield, xii, 122, 127, 130, 133, 134, 135, 136, 137, 138, 139, 140, 141, 143
yield-per-recruit, 139

Z

zooplankton, 56

Related Nova Publications

TELEOSTS: PHYSIOLOGY, EVOLUTION AND CLASSIFICATION

EDITOR: Mikael Herleif

SERIES: Marine and Freshwater Biology

BOOK DESCRIPTION: The authors begin this compilation by analyzing catalase and peroxidase enzymes in different Black Sea teleosts related to their taxonomic, physiological and ecological position and evaluation of the anthropogenic impact on these antioxidant enzymes in fish tissues.

SOFTCOVER ISBN: 978-1-53613-660-9
RETAIL PRICE: $82

MUSSELS: CHARACTERISTICS, BIOLOGY AND CONSERVATION

EDITORS: Brooke Mansom and Ellie Grover

SERIES: Marine Science and Technology

BOOK DESCRIPTION: In this compilation, the authors examine a plant dealing with mussel shell waste as input in its valorisation process, which is considered a priori an eco-friendly solution to the disposal of these products.

HARDCOVER ISBN: 978-1-53613-459-9
RETAIL PRICE: $160

To see complete list of Nova publications, please visit our website at www.novapublishers.com